PPM

Practical Problems in Mathematics

FOR WELDERS

6TH EDITION

This book is dedicated to the future and happiness of my sons Michael and David, the memory of my parents Walter and Helen, and to the memory of Dr. Wilhelm Reich:

"Love, work, and knowledge are the wellsprings of life: they should also govern it."—WR

CONTENTS

SECTION 3 DECIMAL FRACTIONS

SECTION 4 AVERAGES, PERCENTAGES, AND MULTIPLIERS

SECTION 5 METRIC SYSTEM MEASUREMENTS

SECTION 6 COMPUTING GEOMETRIC MEASURE AND SHAPES

PREFACE

Practical Problems in Mathematics for Welders, 6e has been written to help students gain experience and confidence in computing problems that are common in a wide variety of welding applications. Welders rely on math to solve everyday problems, from ordering materials to planning for their economical use. This book will show students examples of the many types of problems that welders encounter. At the same time, it will provide information about a variety of welding careers. Common welding terminology and applications are incorporated throughout the book.

DELMAR'S PPM SERIES

This text is one of a series of workbooks designed to offer students practical problem-solving experience in various occupations. The workbooks take a step-by-step approach to mastering basic math skills. Each workbook includes relevant and easily understood problems in a specific vocational field. The workbooks are suitable for any student from junior high through high school and up to the two-year college level. Each text includes a glossary to help students with technical terms. *Practical Problems in Mathematics for Welders, 6e* includes an Appendix with information on English and SI measurements, important formulas, and Answers to Odd-Numbered Problems. For more information about this series and a current list of titles, please visit www.CengageBrain.com.

SERIES FEATURES

The workbooks in Delmar's PPM series take a step-by-step approach to mastering essential math skills. At the start of each unit, a brief introductory section provides a basic explanation of the concepts necessary to complete the problems in the unit. Examples are presented to help the learner review the mathematical principles. The problems in each unit progress from basic examples of the math concepts to more complex examples that require critical thinking. As students progress through each unit, they will become more proficient at solving a wide variety of math problems.

THIS BOOK'S APPROACH

Practical Problems in Mathematics for Welders, 6e begins with a review of basic operations with whole numbers, fractions, and decimals, and progresses through measurements, area and volume calculations, and angular development, to finish with a section on bends, stretchouts, economical layout, and take-offs. Topical sections are divided into short units to give teachers maximum flexibility in planning and to help students achieve maximum skill mastery. Instrutors may choose to use this book as a stand-alone text or as a supplemental workbook to a theory-based text.

NEW TO THIS EDITION

The sixth edition of *Practical Problems in Mathematics for Welders* is updated to include a new section on choosing and using a calculator, and more information on using common measuring tools for welders, including micrometers, calipers, steel tape, and the steel square. The explanations in Section 7 on angular development and measurement have been clarified, and new practical problems have been added throughout the text.

SUPPLEMENTS

The supplements package for this edition has been revised and expanded to include a new Instructor Resources CD and Applied Math CourseMate, a new on-line tool that can help students and teachers build lasting math skills.

Instructor Resources

The Instructor Resources CD provides the following support for teachers:

- updated answers to all text problems,
- computerized test banks in ExamView® software,
- PowerPoint® presentations, and
- an Image Gallery including all text figures.

Applied Math CourseMate

Every text in Delmar's PPM series includes Applied Math CourseMate, Cengage Learning's on-line solution for building strong math skills. Students and instructors alike will benefit from the following CourseMate Resources:

- an interactive eBook, with highlighting, note-taking, and search capabilities;
- interactive learning tools including:
  - quizzes,
  - flashcards,
  - PowerPoint slides,
  - skill-building games;
- and more!

Instructors will be able to use Applied Math CourseMate to access the Instructor Resources and other classroom management tools. Go to login.cengagebrain.com to access these resources, and look for this icon CourseMate to find resources related to your text in Applied Math CourseMate.

A MESSAGE TO STUDENTS

Metal has been an integral part of human life since the Iron Age and Bronze Age. In the beginning, metal workers were responsible for the shaping and joining of metal for the production of art, weaponry, cookware, and in crude building practices. Today, current industry demands and welders skills are producing the massive superstructures of skyscrapers and aircraft carriers, schools and missile defense systems, oil and gas pipelines aboveground and underwater, choppers, cars and bridges, robotics and power generation, fighter aircraft and the guts of billion-dollar computer chip factories. Welding projects abound in hundreds of applications and in all types of large and small business. Mostly working with steel, and, to a growing extent, plastics and ceramics, today's metalworker is typically a welder. More than 50% of the gross national product of the United States is associated in one way or another with welded and/or bonded products. As the population of the planet grows, so will this trade!

WORK OUTLOOK

Utilizing various methods of joining materials such as the MIG, TIG, arc weld or gas weld processes, the future looks bright for anyone seeking a rewarding career in this field. Job growth in welding is expected to be huge. Employers are already reporting difficulty finding a sufficient number of qualified workers. According to the Federal Bureau of Labor Statistics (BLS), many in the industry are approaching retirement age: the retirement of experienced workers alone opens great opportunities. Coupled with an expected growth in consumer population, industry reports a need for at least 238,700 new and replacement employees across all welding job codes, and an increase of up to 37% of the workforce through 2019. With growth in the demand for products, transportation, defense and construction, welding students of today are in an excellent position to become tomorrow's skilled and richly rewarded workers in high demand.

Salaries for non-union welders range from an entry level $10.00 to $18.00 per hour, up to double that for qualified workers. Many welders belong to unions, such as the Pipefitters, Ironworkers, Sheet Metal Workers, Boilermakers, Iron Ship Builders, Machinists, and Carpenters. Depending on the area of the country in which you work, entry-level apprentices in unions can earn $22.00 to $30.00 per hour including benefits, and up to double that as journeymen. If a welder advances to specialized skills, salaries over $100,000 per year are common. In addition, opportunities for managerial positions, inspectors, engineers, and instructors are available and in demand as well. Learning the skills, however, is critical for all of this to happen. Experience and certification in welding processes is gained in welding schools, apprenticeships, and in the workplace.

Whether you are seeking a new job or trade, welding offers unlimited potential, growth, and security. As a student or apprentice, you are well on your way to success. Work safely, work hard, study diligently, and the future is yours with a great career in welding.

Contact information:

American Welding Society
550 NW LeJeune Rd.
Miami, Florida 33126
1-800-443-9353 or 305-443-9353
www.aws.org
Western Apprenticeship Coordinators Association
www.azwaca.org

ACKNOWLEDGMENTS

Special thanks to Mary Cook, Coordinator at the Arizona Department of Transportation (ADOT) for the statewide Highway Construction program, with whom I've had the pleasure of working since the program's inception in 1996. Mary, the first female journeyman Pipefitter in Arizona, was co-developer of the program along with Paula Goodson, the former Director of the Governors' Division for Women of the State of Arizona.

Further appreciation goes to the special group of men, women, and the organizations they belong to in the Western Apprenticeship Coordinator's Association (WACA), the union apprenticeship programs and Joint Apprenticeship Training Committees (JATC's) including the Operating Engineers, Pipe Fitting Trades, Carpenters, Ironworkers, Bricklayers, Painters, Sheetmetal Workers, Electricians, etc. The quality of their dedication, and their benefit to the community and worker, is enormous. The Highway program could not have been a success had it not been for their dedicated involvement. Special thanks to Ed Yarco, Raul Garcia, John Malmos, Mike Wall, George Facista, Jerry Bellovary, Willie Higgins, Partricio Melivilu, Cheryl Williams, Dennis Anthony, George Sapien, Kritina Mohr, and numerous other participants.

During my work at the Maricopa Skill Center (MSC), a division of Maricopa County Community College District (MCCCD), I was fortunate in being given the position of developing the operation and curriculum, coordinating the responsibilities, and instructing the students of the Highway Construction program since its' inception in 1996. Additional leadership and instructional responsibilities at MSC included the City of Phoenix Home Maintenance program, the Secretary of Labor's Commission on Achieving Necessary Skills (SCANS), and the OSHA-Compliance Safety Team. I taught math to various vocational groups including the Welding, Highway Construction, Machine Trades, Meat Cutting, Culinary, and Auto Body departments, as well as blueprint reading and OSHA/SSTA standards. Thanks go to all the highly qualified and caring personnel at our school.

The Highway Construction program became a leader in the nation for Federal Highway Administration training programs, and received recognition from Arizona Governor Jane Dee Hull. This was made possible only through the combined dedication of all involved: co-workers, the apprenticeship JATC's, and numerous partners throughout the community. It is a privilege to have worked with such terrific men and women. The difference they make in people's lives and the personal and professional relationships established make me proud and grateful to each and every one who was a part of these efforts. My heartfelt appreciation goes to them all.

Thank you also to Ann C. Lavit, Ph.D., for the encouragement to enter the education field, to all those who made my work possible including Stanley Grossman, former Director of MSC and in memorium to Susan McRae and Oscar Gibbons, former Assistant Directors of MSC, and to my son Michael Chasan, copywriter, for the main body of A Message to Students.

I wish to acknowledge a special appreciation to all my students: though you thanked me for my teachings, I in turn give thanks for all that you taught me. Your courage and desire to succeed, many of you overcoming difficult personal situations, was inspiring. I felt honored to be a part of your journey. Thank you.

The author and Delmar, Cengage Learning wish to acknowledge and thank the review panel for the many suggestions and comments during development of this edition. A special thanks to Mike Standifird for preparation of the testbank. Members of the review panel include:

James Mike Daniel
Peninsula College
Port Angeles, WA

Robert Dubuc
New England Institute of Technology
Warwick, RI

Jerry Galyen
Pinellas Technical Education Center
Clearwater, FL

Jimmy Kee
Tennessee Technology Center
McKenzie, TN

Barnabie Mejia
Western Technical College
El Paso, TX

Mike Standifird
Angelina College
Lufkin, TX

Dale Szabla
Anoka Technical College
Anoka, MN

Steve White
Mobile Technical Institute
Mobile, AL

ABOUT THE AUTHOR

Robert Chasan taught math, highway construction, and OSHA safety for the Pre-Apprenticeship Training in Highway Construction Program at the Maricopa County Community College District in Phoenix, Arizona. Under Professor Chasan's tutelage, this program became a national leader funded by the Federal Highway Administration. Mr. Chasan also coordinated his school's SCANS Lab (Secretary of Labor's Commission on Acquiring Necessary Skills) teaching math, blueprint reading, and SCANS skills to the welding and machine trades students and the school's other trades programs. Professor Chasan led the school Safety Team and taught OSHA-10, OSHA-30, and SSTA-16. In addition, Mr. Chasan coordinated the Future Builders' Academy, a highly supportive high school student program sponsored by the Arizona Builders' Alliance in Phoenix.

A certified Arizona Community College instructor, Professor Chasan is well recognized for his vocational instruction. Mr. Chasan resides in Mesa, Arizona.

SECTION 1

WHOLE NUMBERS

UNIT 1

Addition of Whole Numbers

Basic Principles

The addition of whole numbers is a procedure necessary for all welders to use.

Each whole number has a decimal point (.) at the end. The decimal point is not usually needed or used until decimal fractions are calculated.

EXAMPLE: 647 is a whole number. It can be written with the decimal point or without it. The decimal point, visible or not, is always at the end of a whole number.

647 = 647.

RULE: Starting from the right side of any whole number:

the first number on the right is named "ones";

the second number from the right is named "tens"; and

the third number from the right is named "hundreds."

Place names continue in like fashion as more numbers are added.

EXAMPLE: 647

Hundreds	Tens	Ones
6	4	7

When adding whole numbers, the following applies:
Whole numbers 1 through 9 (ones) are placed one beneath the other in a column. The plus sign (+) is used.

EXAMPLE: 3 + 6 + 9

$$\begin{array}{r} 3 \\ 6 \\ +\ 9 \\ \hline 18 \end{array}$$

When whole numbers greater than 9 are added, the numbers are arranged starting with the ones lined up beneath each other. Ones are lined up in their own column, tens are in their own column, etc. Addition and subtraction calculations (see Unit 2) start with the ones column.

EXAMPLE: 3 + 6 + 9 + 14 + 214

Setup	Step 1	Step 2	Step 3
	2	2	2
3	3	3	3
6	6	6	6
9	9	9	9
14	4	14	14
+ 214	+ 4	+ 14	+ 214
	6	46	246

STEP 1 The ones column adds up to 26. The 6 is placed beneath the ones column, and the 2 is carried over to the tens column.

STEP 2 The tens column, including the 2 that was carried over, adds up to 4. Place the 4 under the tens column.

STEP 3 The hundreds column adds up to 2. Place the 2 under the hundreds column.

Practical Problems

1. An inventory of steel angle in four separate areas of a welding shop lists 98 feet, 47 feet, 221 feet, and 12 feet. Find the total amount of steel angle in inventory. ______

2. Layout work for a welded steel bar is shown. Determine the total number of inches of steel used in the length of the bar. ______

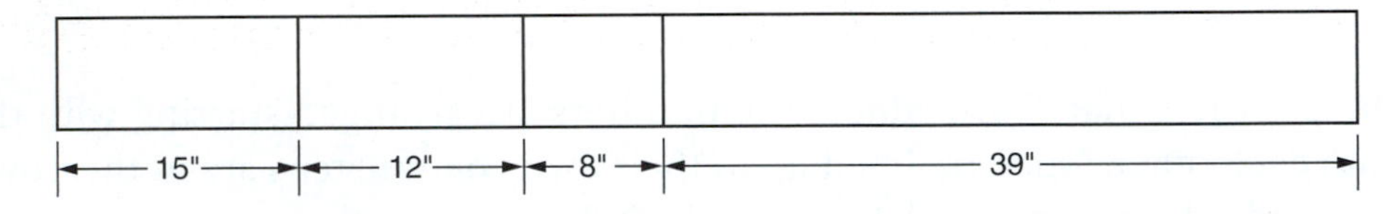

3. A welded steel framework consists of: plate steel, 1,098 pounds; key stock, 13 pounds; bolt stock, 98 pounds; and channel, 822 pounds.

   What is the total weight of steel in the framework? ______

UNIT 2

Subtraction of Whole Numbers

Basic Principles

It is often necessary for welders to be able to subtract (take away) one amount from another. The minus sign (−) is used to indicate subtraction.

EXAMPLE 1: The shop in which you are working needs to fabricate 15 plates for a customer. If 9 have already been made, how many more are needed? (What is 15 minus 9?)

STEP 1 The ones are lined up beneath each other, as in addition. Subtraction is started with the ones column.

$$\begin{array}{r} 15 \\ -\ 9 \\ \hline \end{array}$$

STEP 2 The answer is placed beneath the ones column.

$$\begin{array}{r} 15 \\ -\ 9 \\ \hline 6 \end{array}$$

 There are 6 more plates that need to be made.

Borrowing to Solve

EXAMPLE 2: 234 − 76 =

NOTE: Dashed lines show separated ones, tens, and hundreds columns.

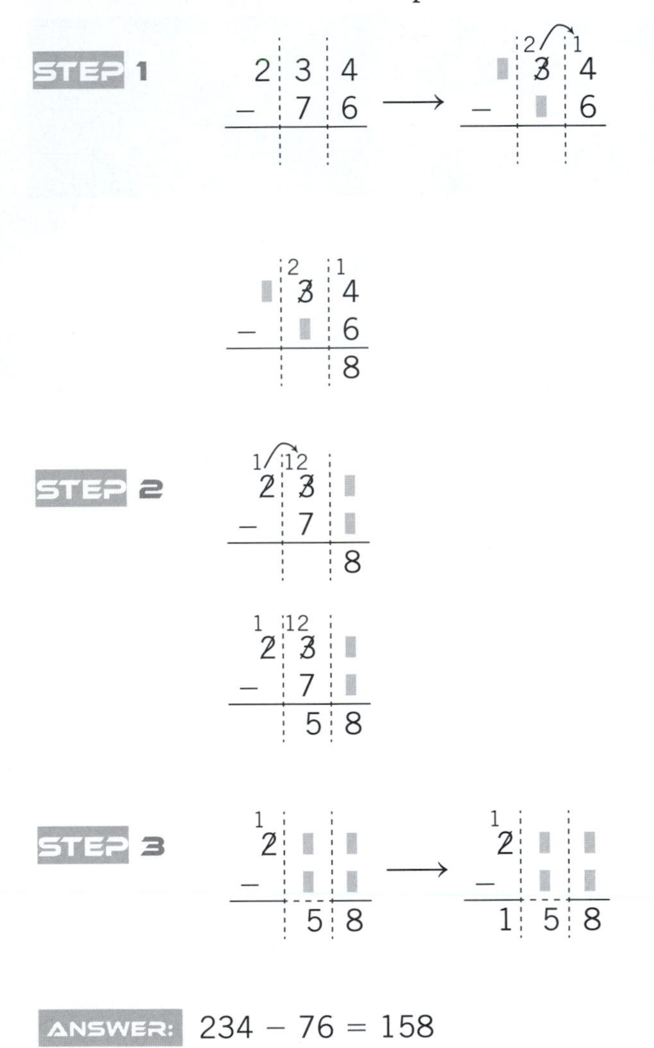

STEP 1

Calculations are started in the ones column.

6 cannot be subtracted from 4. Borrow 1 from the 3 in the tens column and place in front of the 4. Because the 1 came from the tens column, you now have 14 instead of 4.

14 − 6 = 8

STEP 2

Seven cannot be subtracted from 2. Borrow 1 from the hundreds column on the left and place in front of the 2. This now gives you 12.

12 − 7 = 5

STEP 3

Continue subtraction in the hundreds columns.

ANSWER: 234 − 76 = 158

Practical Problems

1. A welder is required to shear-cut a piece of sheet steel as shown in the illustration. After the cut piece is removed, how much sheet, in inches, remains from the original piece? __________

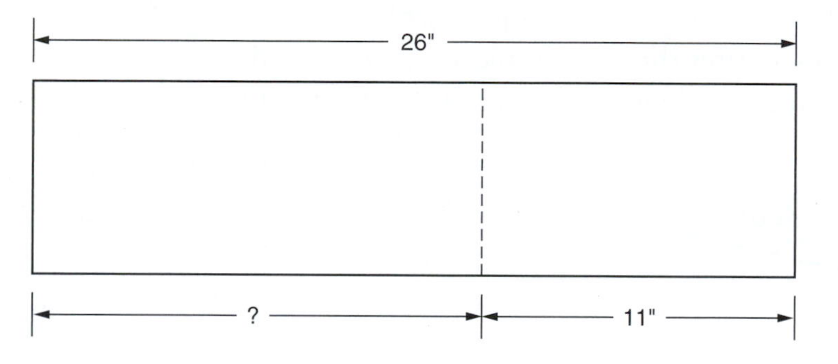

2. The inventory of a scrap pile is as follows:

   channel iron, 89″
   plate steel, 134″
   key stock, 42″
   pipe, 65″
   flat iron strap, 184″

   A pipe support is welded using the following materials from the inventory:

   channel iron, 61″
   plate steel, 106″
   key stock, 39″
   pipe, 22″
   flat iron strap, 73″

   What is the balance remaining of each item, in inches?

   Channel iron a. __________

   Plate steel b. __________

Key stock c. ______________

Pipe d. ______________

Flat iron strap e. ______________

3. A length of pipe is cut as shown. After the two cut pieces are removed, how long, in inches, is the remaining length of pipe? Disregard waste caused by the width of the cut. ______________

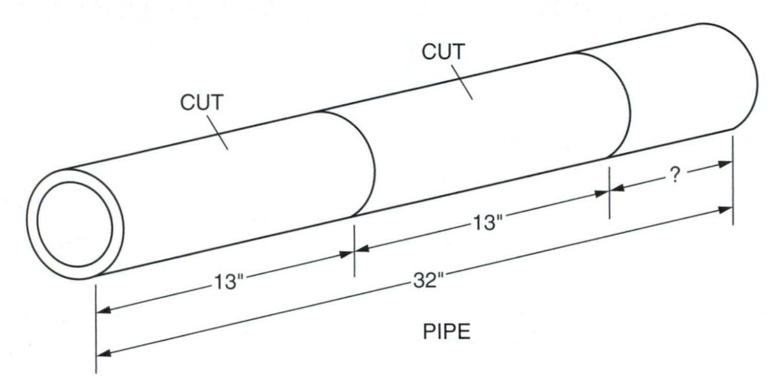

4. A stock room has 18,903 pounds of plate steel. A steel storage tank is welded from 1,366 pounds of that plate. How many pounds of plate remain in stock? ______________

UNIT 3

Multiplication of Whole Numbers

Basic Principles

Multiplication builds on the principles of addition: it is helpful in figuring large quantities quickly when addition may be too slow.

The (×) symbol is used to show multiplication, although other ways to show multiplication will be taught later in the book.

Each part of the problem has a name:

The top number, the one being multiplied, is called the multiplicand.

The lower number, the one doing the multiplying, is called the multiplier.

The multiplication tables are used often in multiplication. Use the tables found in this unit to refresh your memory. If you don't know the tables, use this unit to begin memorizing them. If you spend the time now to memorize each table, one at a time, all math will be a bit easier for you to handle. A good way to memorize the tables is to practice a little every day, repeating the numbers out loud with a friend or to yourself.

Procedures in multiplication:

EXAMPLE 1: How many electrodes are available for use if a welder has 3 bundles, each containing 26 electrodes? ($26 \times 3 = ?$)

STEP 1 Set up the problem:

$$\begin{array}{rl} 26 & \text{(multiplicand)} \\ \underline{\times\ 3} & \text{(multiplier)} \\ \multicolumn{2}{l}{\text{(product = answer)}} \end{array}$$

STEP 2 Multiply the ones column first ($3 \times 6 = 18$). The 8 is placed underneath the ones column, and the 1 is carried over to the top of the tens column.

$$\begin{array}{r} 6 \\ \underline{\times\ 3} \end{array} \qquad \begin{array}{r} {\scriptstyle 1} \\ 6 \\ \underline{\times\ 3} \\ 8 \end{array}$$

STEP 3 Next, multiply the tens column ($3 \times 2 = 6$). Add the number carried over 1. ($6 + 1 = 7$). Place the 7 under the tens column.

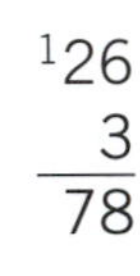

$$\begin{array}{r} {}^{1}26 \\ \underline{3} \\ 78 \end{array}$$

ANSWER: There are 78 electrodes available.

EXAMPLE 2: 238×47

STEP 1

$$\begin{array}{r} 238 \\ \underline{\times\ 47} \end{array}$$

STEP 2 Start multiplication with the ones column ($7 \times 8 = 56$). Place the 6 as shown, carry over the 5 for the next step.

$$\begin{array}{r} {\scriptstyle (5)}\ \ \\ 238 \\ \underline{\times\ \ 7} \\ 6 \end{array}$$

STEP 3 Multiply the next number, and continue until the series is completed. $(7 \times 3 = 21, 21 + 5 = 26)$ $(7 \times 2 = 14, 14 + 2 = 16)$.

```
   5       25       25
  238      238      238
× ▮7     × ▮7     × ▮7
-----    -----    -----
    6       66     1666
```

STEP 4 Multiply with the second number in the multiplier, and continue until the series is completed.

```
            5        13        13
  238      238      238       238
× 4▮     × 4▮     × 4▮      × 4▮
-----    -----    -----     -----
         ▮▮▮▮     ▮▮▮▮      ▮▮▮▮
            2       52      952
```

STEP 4a Multiplication is complete.

```
  238      25        13
× 47       238       238
-----    × 47      × 47
         -----     -----
          1666      1666
                    952
                   -----
```

STEP 5 Add the columns.

```
  ¹1666
 + 952
 ------
  11186
```

ANSWER: $238 \times 47 = 11{,}186$

Practical Problems

1. A welded support is illustrated.

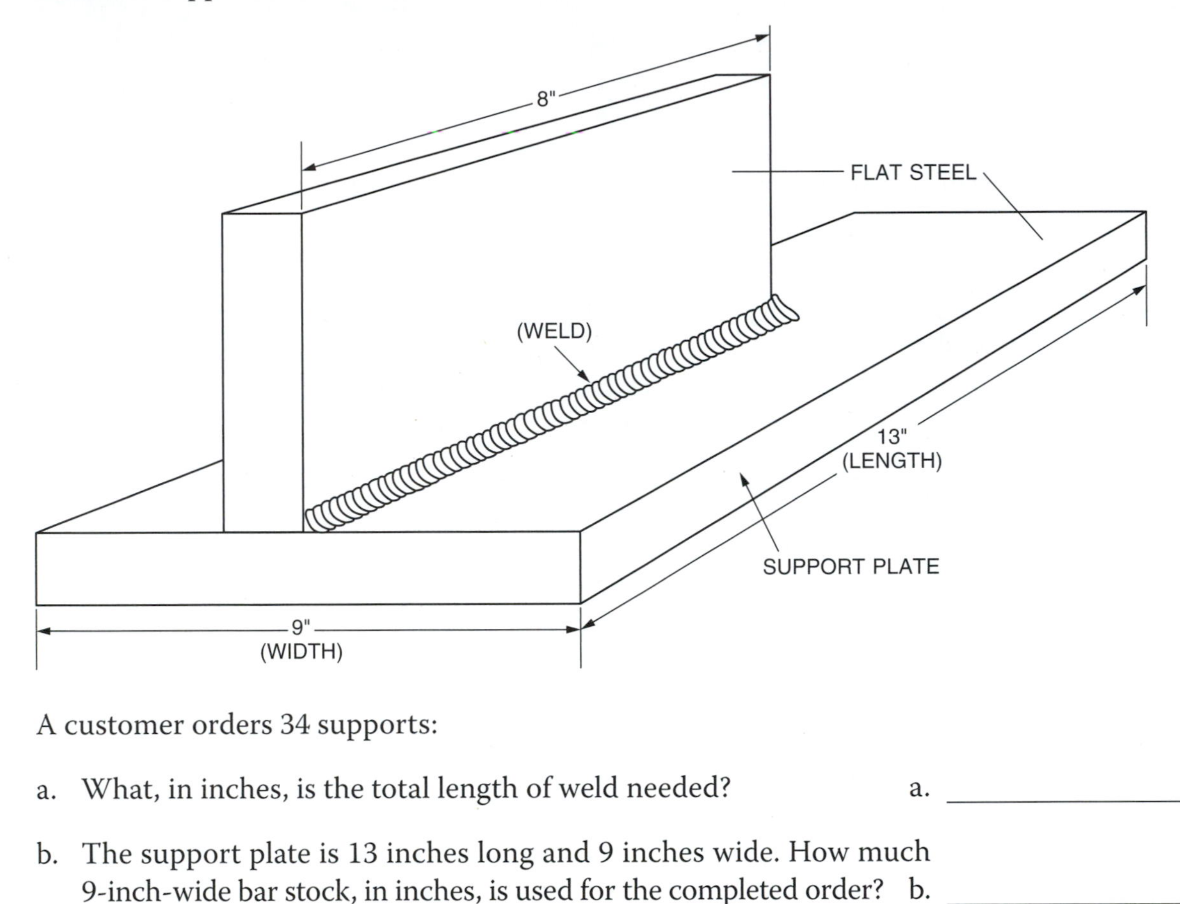

A customer orders 34 supports:

a. What, in inches, is the total length of weld needed? a. ______

b. The support plate is 13 inches long and 9 inches wide. How much 9-inch-wide bar stock, in inches, is used for the completed order? b. ______

c. Each support weighs 14 pounds. What is the weight in pounds of the total order? c. ______

2. A welded tank support requires 14 pieces of wide-flange beam to be cut. Each piece of beam is 33 inches long. What is the total number of inches of beam used? Disregard waste caused by the width of any cut. ______

3. A job requires 1,098 pieces of bar stock, each 9 inches long. What is the total length of bar stock required, in inches? ______

4. A welder tack welds 215 linear feet of steel support columns per hour. How many feet of support columns are completed in an 8-hour shift? __________

5. A MIG welding unit is run at a wire feed speed of 30″ per minute. How many inches of wire are used in an hour? In 8 hours of continuous run? __________ __________

6. A pump moves 56 gallons of water per second. How many gallons are moved in an hour? In a week of continuous run? __________ __________

MULTIPLICATION TABLES

× 2	× 3	× 4	× 5	× 6
$2 \times 1 = 2$	$3 \times 1 = 3$	$4 \times 1 = 4$	$5 \times 1 = 5$	$6 \times 1 = 6$
$2 \times 2 = 4$	$3 \times 2 = 6$	$4 \times 2 = 8$	$5 \times 2 = 10$	$6 \times 2 = 12$
$2 \times 3 = 6$	$3 \times 3 = 9$	$4 \times 3 = 12$	$5 \times 3 = 15$	$6 \times 3 = 18$
$2 \times 4 = 8$	$3 \times 4 = 12$	$4 \times 4 = 16$	$5 \times 4 = 20$	$6 \times 4 = 24$
$2 \times 5 = 10$	$3 \times 5 = 15$	$4 \times 5 = 20$	$5 \times 5 = 25$	$6 \times 5 = 30$
$2 \times 6 = 12$	$3 \times 6 = 18$	$4 \times 6 = 24$	$5 \times 6 = 30$	$6 \times 6 = 36$
$2 \times 7 = 14$	$3 \times 7 = 21$	$4 \times 7 = 28$	$5 \times 7 = 35$	$6 \times 7 = 42$
$2 \times 8 = 16$	$3 \times 8 = 24$	$4 \times 8 = 32$	$5 \times 8 = 40$	$6 \times 8 = 48$
$2 \times 9 = 18$	$3 \times 9 = 27$	$4 \times 9 = 36$	$5 \times 9 = 45$	$6 \times 9 = 54$
$2 \times 10 = 20$	$3 \times 10 = 30$	$4 \times 10 = 40$	$5 \times 10 = 50$	$6 \times 10 = 60$

× 7	× 8	× 9	× 10
$7 \times 1 = 7$	$8 \times 1 = 8$	$9 \times 1 = 9$	$10 \times 1 = 10$
$7 \times 2 = 14$	$8 \times 2 = 16$	$9 \times 2 = 18$	$10 \times 2 = 20$
$7 \times 3 = 21$	$8 \times 3 = 24$	$9 \times 3 = 27$	$10 \times 3 = 30$
$7 \times 4 = 28$	$8 \times 4 = 32$	$9 \times 4 = 36$	$10 \times 4 = 40$
$7 \times 5 = 35$	$8 \times 5 = 40$	$9 \times 5 = 45$	$10 \times 5 = 50$
$7 \times 6 = 42$	$8 \times 6 = 48$	$9 \times 6 = 54$	$10 \times 6 = 60$
$7 \times 7 = 49$	$8 \times 7 = 56$	$9 \times 7 = 63$	$10 \times 7 = 70$
$7 \times 8 = 56$	$8 \times 8 = 64$	$9 \times 8 = 72$	$10 \times 8 = 80$
$7 \times 9 = 63$	$8 \times 9 = 72$	$9 \times 9 = 81$	$10 \times 9 = 90$
$7 \times 10 = 70$	$8 \times 10 = 80$	$9 \times 10 = 90$	$10 \times 10 = 100$

MULTIPLICATION TABLES

1	**2**	**3**	**(4)**	**5**	**(6)**	**7**	**8**	**9**	**10**
2	4								
3	6	9							
4	8	12	16						
5	10	15	20	25					
(6)	12	18	[24]	30	36				
7	14	21	28	35	42	49			
8	16	24	32	40	48	56	64		
(9)	18	27	36	45	[54]	63	72	81	
10	20	30	40	50	60	70	80	90	100

Examples: $6 \times 4 = 24$
$9 \times 6 = 54$

UNIT 4

Division of Whole Numbers

Basic Principles

Division is a method used to determine the quantity of groups available in a given number. Several symbols can be used to show division:

÷ indicates "divided by."

The line in a fraction ($\frac{}{}$) is called a fraction line and also indicates "divided by." Fractions will be taught in the following section.

The division box, $\overline{)}$, is used to do the actual work of dividing.

The number outside of the box, which is the number doing the dividing, is called the divisor.

The number inside the box, which is the number to be divided, is called the dividend.

EXAMPLE 1: Seven welders are assigned the job of fabricating a steel platform. They are given 217 electrodes. Dividing the electrodes equally, how many electrodes does each welder get?

PROBLEM: $217 \div 7 = ?$

Setup	Step 1	Step 2	Step 3
$7\overline{)217}$	$7\overline{)2\ \ }$	$\begin{array}{r} 3 \\ 7\overline{)21\ } \\ -21 \\ \hline 0 \end{array}$	$\begin{array}{r} 31 \\ 7\overline{)217} \\ -21\downarrow \\ \hline 07 \\ -7 \\ \hline 0 \end{array}$

STEP 1 7 into 2 will not go. There are no groups of 7 in 2.

STEP 2 7 into 21 goes 3 times. There are 3 groups of 7 in 21.

Bring down last number.

STEP 3 7 into 7 goes once. There is 1 group of 7 in 7.

ANSWER: $217 \div 7 = 31$

Each welder will get 31 electrodes.

EXAMPLE 2: A worker bolts plates onto a frame support. Each plate needs 16 bolts. If there are 83 bolts available, how many plates can be fully bolted?

(A group of 16 bolts will finish 1 plate. How many groups of 16 are there in 83?)

STEPS: $83 \div 16 = ?$

$$16\overline{)83} \qquad \begin{array}{r} 5 \\ 16\overline{)83} \\ -80 \\ \hline 3 \end{array} \text{ left over}$$

ANSWER: There are 5 groups of 16. Even though 3 bolts are left over, only 5 plates can be fully bolted.

Practical Problems

1. A 16′ length of I-beam is in stock. How many full 3′ lengths of I-beam can be cut from this piece? Disregard waste caused by the width of the cuts. __________

2. A steel support has 4 holes punched at equal distance from each other. Find, in inches, the center-to-center distance between the holes. __________

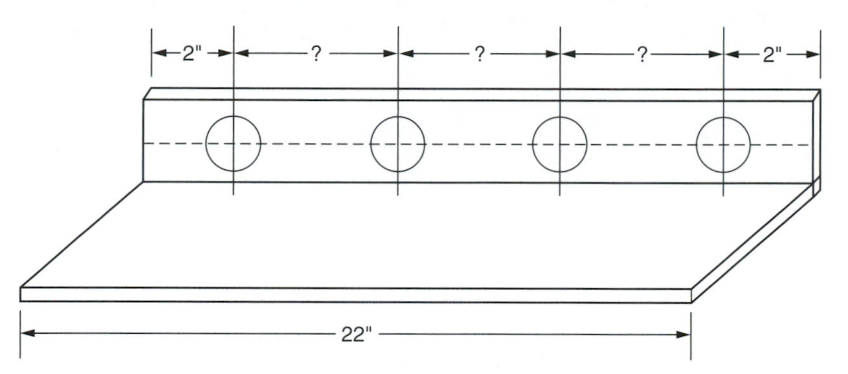

3. A welding rod container holds 50 pounds of rod. A welder can burn 4 pounds of rod per hour. How many full hours will the container last? __________

4. A welder can burn 32 pounds of MIG wire in an 8-hour day. How many pounds is the welder using per hour? __________

5a. 35 connecting plates weigh a total of 7,000 pounds. How much does each plate weigh? __________

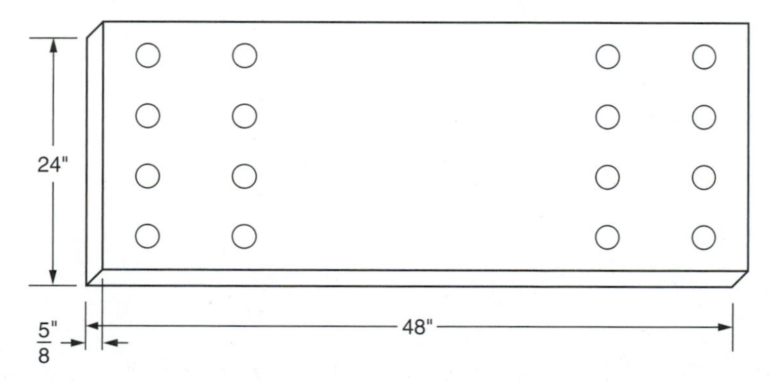

b. If a customer pays a total of $2,625.00 for all 35 plates, how much does each cost? __________

SECTION 2

COMMON FRACTIONS

UNIT 5

Introduction to Common Fractions

Definition of Fractions

There are two types of fractions, both of which describe less than a whole object. The object can be an inch, a foot, a mile, a ton, a bundle of weld rods, other measurements, etc. The two types of fractions are

1. common fractions (fractions) and
2. decimal fractions (decimals).

Common fraction examples are $\frac{1}{2}, \frac{3}{4}, \frac{5}{8}$

Decimal fraction examples are: .50 .75 .625

We will work with fractions (common fractions) in this section. Decimal fractions will be discussed in Section 3.

Basic Principles

The bottom number (the denominator) of every fraction shows the number of pieces any one whole object is divided into; all pieces are of equal size. The top number (the numerator) shows information about the pieces of that divided object.

EXAMPLE: $\frac{3}{8}$

3 is the numerator, and 8 is the denominator. This fraction shows that an object has been divided into 8 equal pieces and that 3 of those 8 pieces are shaded.

Let us work with other simple examples.

If we have one whole unsliced pizza, we can divide it into pieces, and then make fractions about the pizza. This example is cut into 4 pieces (quarters). Fractions concerning this pizza will have the bottom number 4. To describe 1 of those pieces, the fraction is written ¼ (1 of 4 pieces).

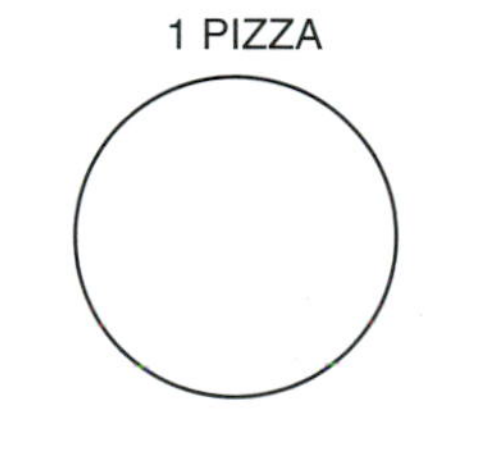

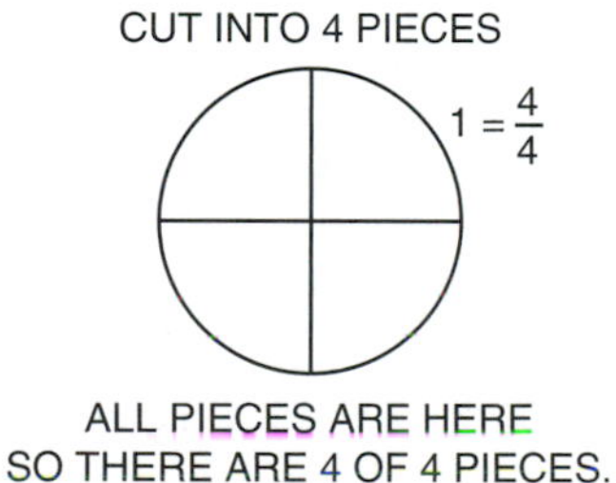

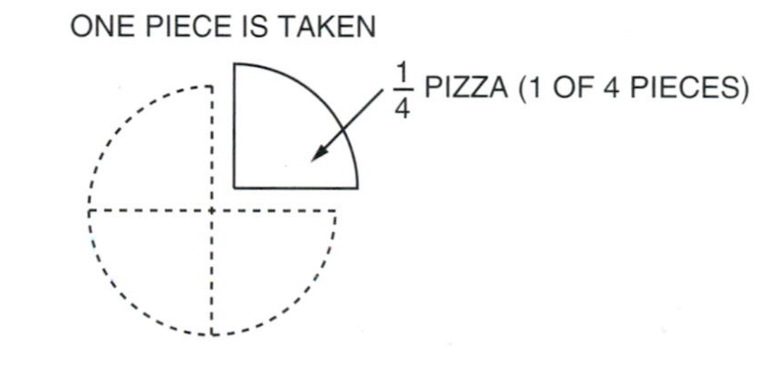

¾ indicates that 3 of 4 pieces are still available.

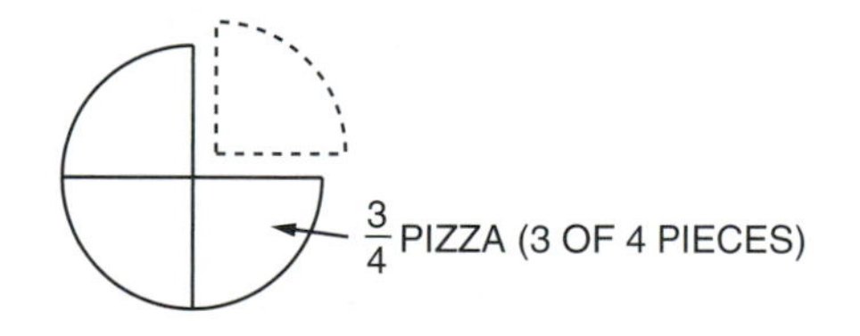

Visualizing Fractions

The proper fraction ¾ represents real slices of pizza. We can draw a picture as follows:

EXAMPLE 1:

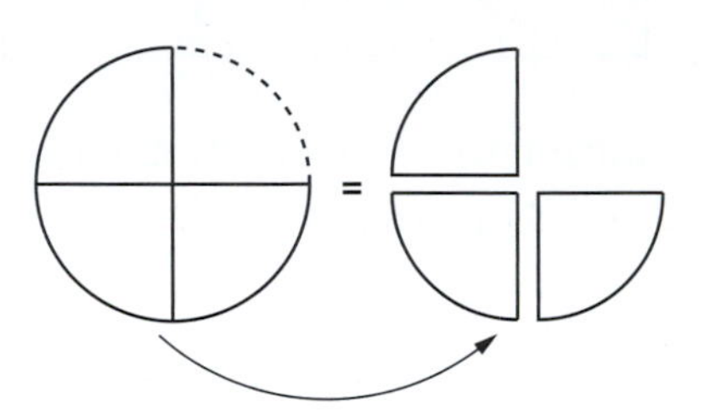

You can see that ¾ pizza is less than 1 whole pizza.

EXAMPLE 2:

If the pizza is cut into 16 pieces, 16 will be the bottom number.

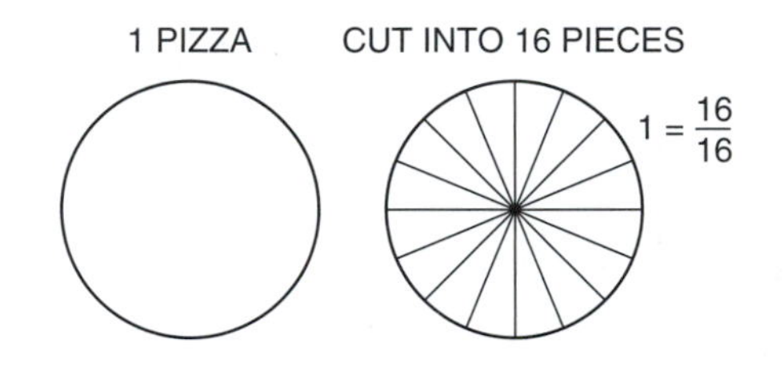

To describe 3 of the pieces, the fraction is written 3⁄16 (3 of 16).

To describe 7 of the pieces, the fraction is written 7⁄16 (7 of 16).

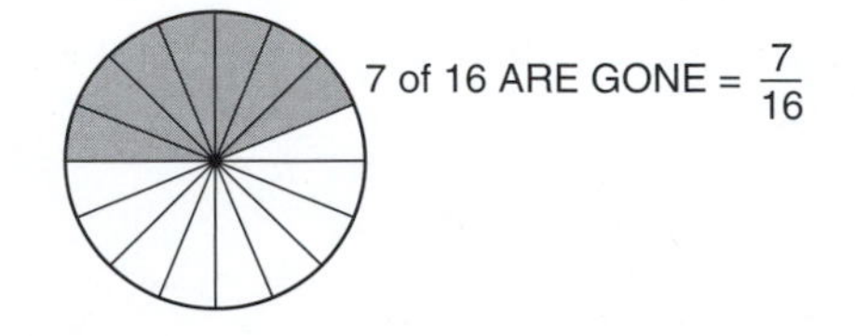

To describe 15 of the pieces, the fraction is written $^{15}/_{16}$ (15 of 16).

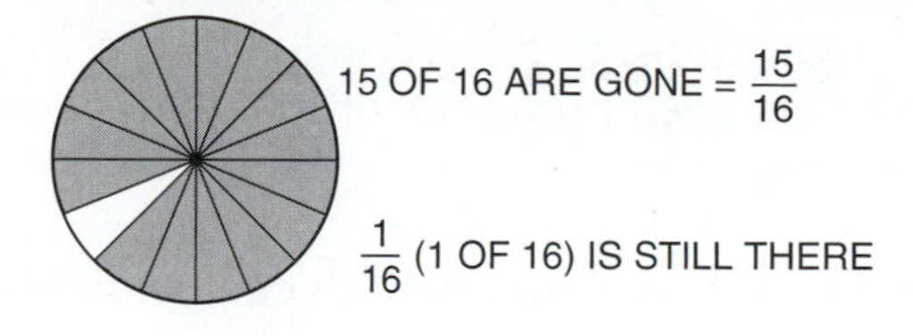

Reducing Fractions

The final step to do to the answer is to reduce the fraction, if possible, to its lowest terms.

EXAMPLE 1:

$$\frac{2}{4} \rightarrow \frac{2 \div 2}{4 \div 2} = \frac{1}{2}$$

$^{2}/_{4}$ is reduced to $^{1}/_{2}$.

EXAMPLE 2:

$$\frac{6}{8} \rightarrow \frac{6 \div 2}{8 \div 2} = \frac{3}{4}$$

$^{6}/_{8}$ is reduced to $^{3}/_{4}$.

EXAMPLE 3:

$$\frac{6}{9} \rightarrow \frac{6 \div 3}{9 \div 3} = \frac{2}{3}$$

$^{6}/_{9}$ is reduced to $^{2}/_{3}$.

Both top and bottom number have to be divided by the same number. In examples 1 and 2, dividing by 2 worked. In example 3, dividing by 3 worked. Some fractions cannot be reduced.

Guide for reducing fractions:

If both the top and bottom are even numbers, or end with even numbers, both can be divided by 2. Examples: $\frac{2}{8}$, $\frac{24}{38}$

If the top and bottom end with 5, or 5 and 0, both can be divided by 5. Examples: $\frac{15}{20}$, $\frac{35}{55}$

If the top and bottom end with 0, both can be divided by 10. Examples: $\frac{30}{70}$, $\frac{110}{330}$

If neither of the above guides work for a particular fraction, experiment by dividing with 3, then 4, then 6, and so on.

Some fractions cannot be reduced, for example, when the top and bottom are 1 number apart (ex: $\frac{3}{4}$, $\frac{7}{8}$, and $\frac{15}{16}$).

We can make fractions that describe part of any object, whether it is part of an inch, part of a foot or mile, part of a pound or ton, and so on.

EXAMPLE 1:

$\frac{5}{8}''$ (five-eights inch) shows that an inch is divided into 8 parts and that 5 of those 8 parts have been measured. The measurement from 0 to a. is $\frac{5}{8}''$.

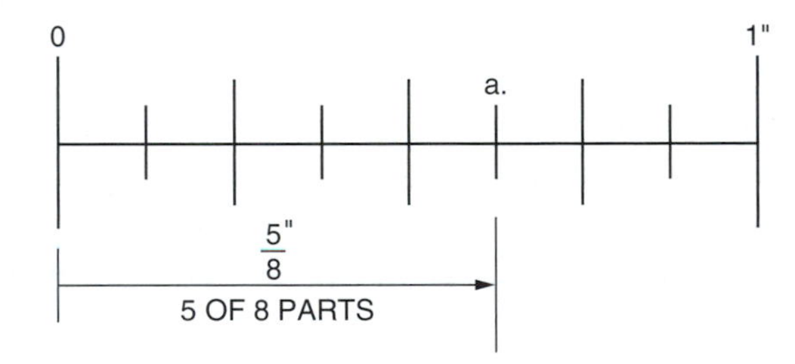

EXAMPLE 2:

7⁄10 of a mile (seven-tenths mile) shows that a mile is divided into 10 parts, and we have measured 7 of those 10 parts. The distance from 0 to b. is 7⁄10 mile.

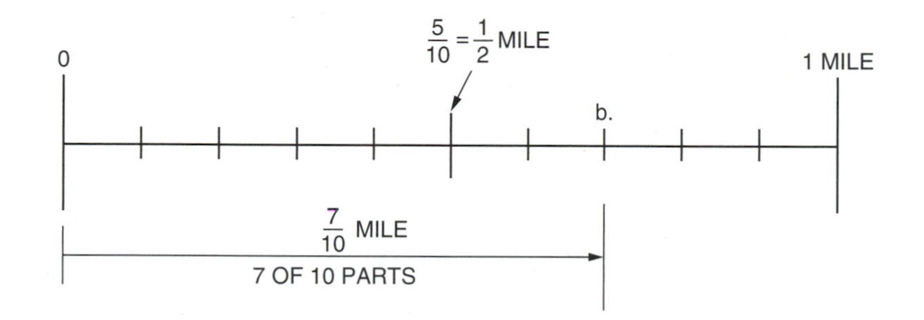

EXAMPLE 3:

¾ ton (three-fourths or three-quarters of a ton) shows that a ton of hay (2,000 pounds) has been divided into 4 parts, and 3 of those 4 parts can be hauled on a flat-bed truck.

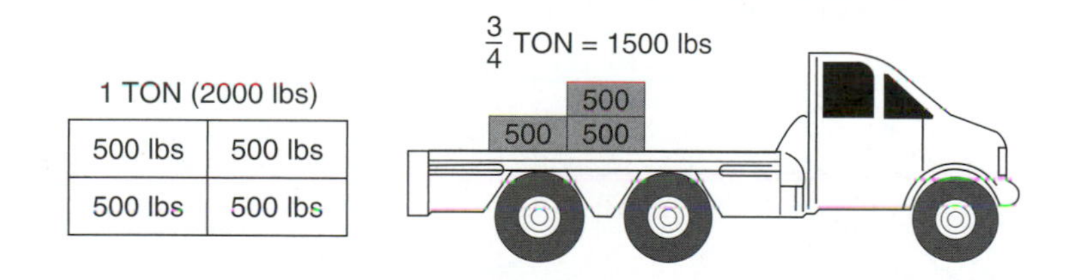

The fractions 5⁄8, 7⁄10, and ¾ and their verbal descriptions "5 of 8 pieces," "7 of 10 parts," and "3 of 4 parts," give your mind a clear picture of each object, how many pieces it was cut up into, and how many of those pieces are being described. With this, you can give accurate information to anyone: a customer, a fellow worker, your foreman, or on a test you may be taking to get into an apprenticeship.

Practical Problems

Make fractions out of the following information; reduce, if possible.

1. 1 foot is divided into 12 inches. Make a fraction of the distance from 0 to a – d

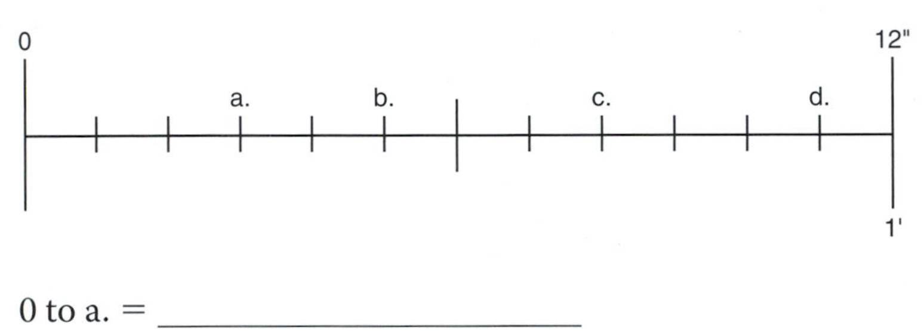

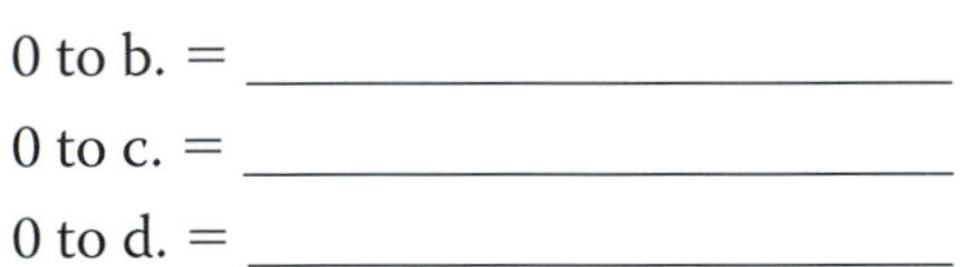

0 to a. = ____________

0 to b. = ____________

0 to c. = ____________

0 to d. = ____________

2. Parts of a bundle of 25 weld rods as fractions

7 rods ____________

10 rods ____________

19 rods ____________

24 rods ____________

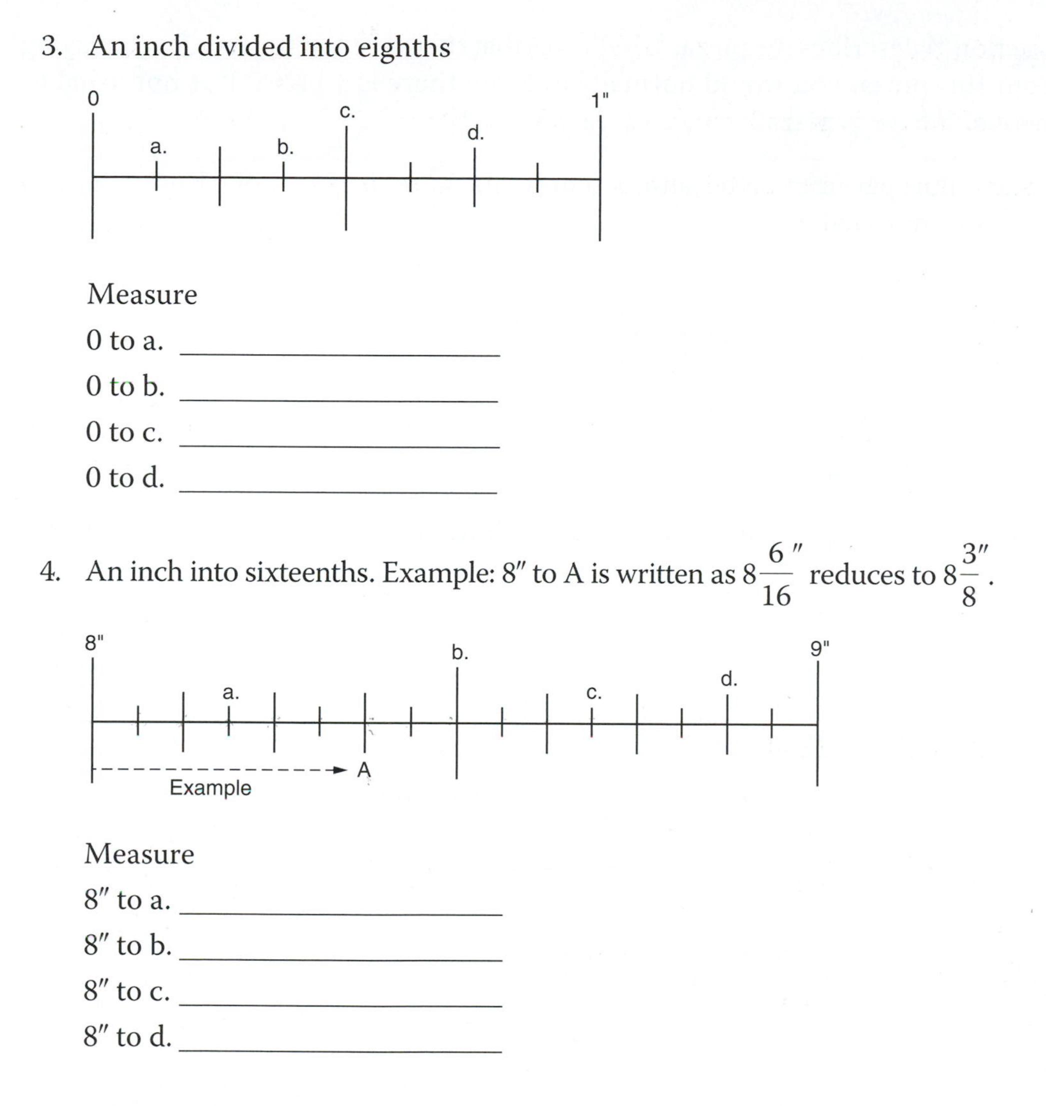

3. An inch divided into eighths

Measure

0 to a. ____________

0 to b. ____________

0 to c. ____________

0 to d. ____________

4. An inch into sixteenths. Example: 8″ to A is written as $8\frac{6}{16}''$ reduces to $8\frac{3}{8}''$.

Measure

8″ to a. ____________

8″ to b. ____________

8″ to c. ____________

8″ to d. ____________

Improper Fractions

All of the fractions we have worked with are called proper fractions because they properly show less than a whole object. However, a fraction is called “improper” when it represents a whole object or more. An example of this can be understood by visualizing another pizza cut into 4 pieces.

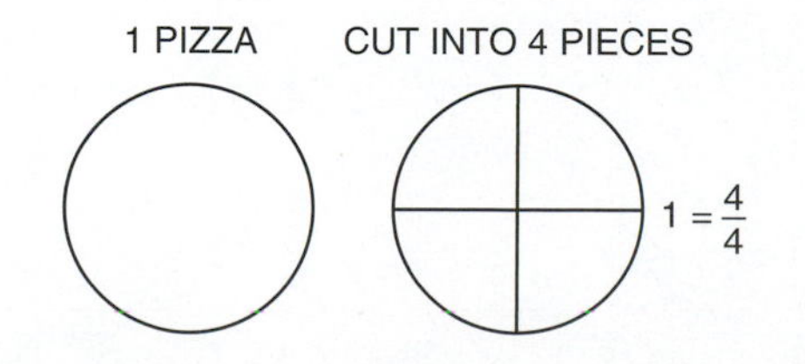

The improper fraction $\frac{4}{4}$ describes the pizza. $\frac{4}{4}$ indicates that there are 4 of 4 pieces. In conveying information about this pizza, you would normally indicate there is 1 pizza. It is not usual to describe to someone, "I have $\frac{4}{4}$ pizza."

The fraction $\frac{7}{4}$ is an improper fraction because it represents more than 1 whole. Here is how to put $\frac{7}{4}$ into its proper form visually.

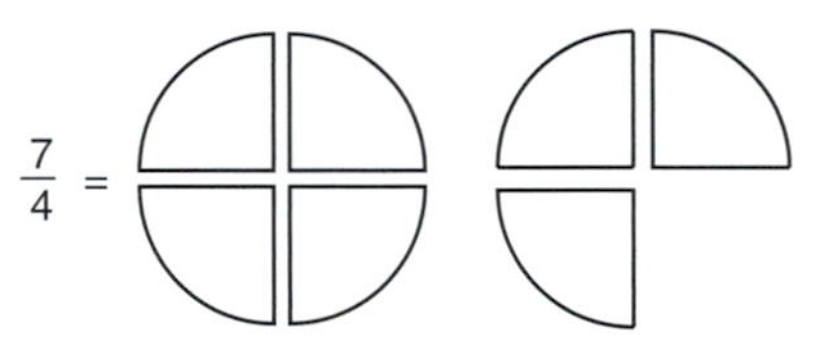

We see 7 quarter pieces of pizza.

Four fourths (four quarters) put together equals 1 whole pizza. In total, we have 1 and $\frac{3}{4}$ pizzas, $1\frac{3}{4}$.

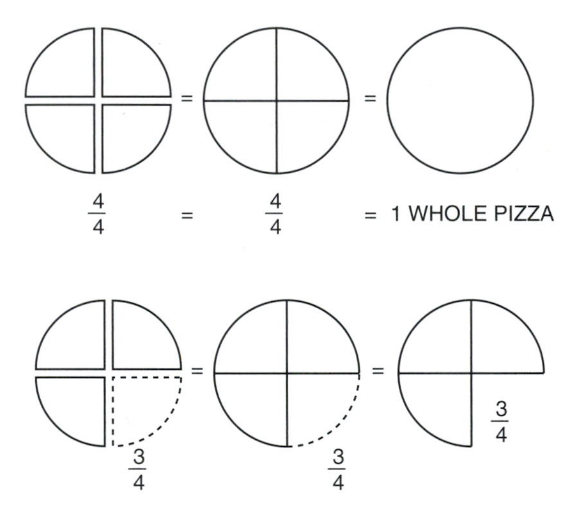

You can determine that a fraction is improper if the top number is the same as the bottom number or if the top number is larger. These fractions convey the picture of a whole object or more. Examples of improper fractions are

$$\frac{8}{8} \quad \frac{5}{2} \quad \frac{16}{15} \quad \frac{11}{8}$$

If there is an improper fraction in the answer to a fraction problem, that improper fraction is to be changed into its proper form which is either a whole or a mixed number. Mathematically, this is done in one step by dividing the bottom number into the top number.

EXAMPLE 1: $\frac{8}{8}$

$$\frac{8}{8} \rightarrow 8\overset{1}{\overline{)8}} = 1 \qquad \frac{8}{8} = 1$$

EXAMPLE 2: $\frac{5}{2}$

$$\frac{5}{2} \rightarrow 2\overline{)5} \rightarrow 2\overset{2}{\overline{)5}} \rightarrow 2\overset{2}{\overline{)5}} = 2\frac{1}{2}$$

$$\begin{array}{r} -4 \\ \hline \end{array} \qquad \begin{array}{r} -4 \\ \hline 1 \end{array}$$

$$\frac{5}{2} = 2\frac{1}{2}$$

Exercises

Decide which fractions are improper and change those fractions to their proper form. Work each problem mathematically. If a fraction is proper, write the word "proper" in the space.

1. $\frac{4}{3}$ ____________
2. $\frac{7}{2}$ ____________
3. $\frac{16}{16}$ ____________
4. $\frac{5}{8}$ ____________

UNIT 6

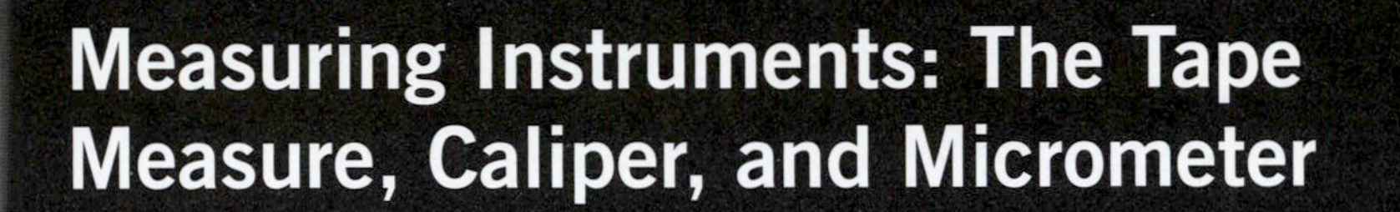

Measuring Instruments: The Tape Measure, Caliper, and Micrometer

Basic Principles

Measuring tapes and rulers are some of the most important tools of the welder: learning to read them accurately is critical and becomes easier with practice.

NOTE: Use this information for Problems 1–6.

© Courtesy of Stanley Tools

Each inch on the tape measure is marked in graduating fractions as shown in illustrations a–f.

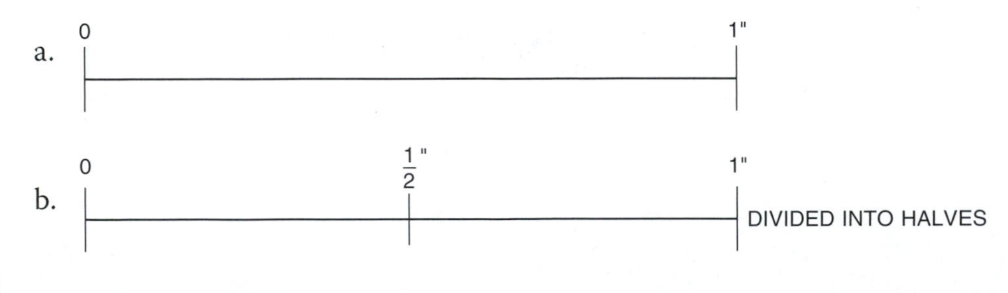

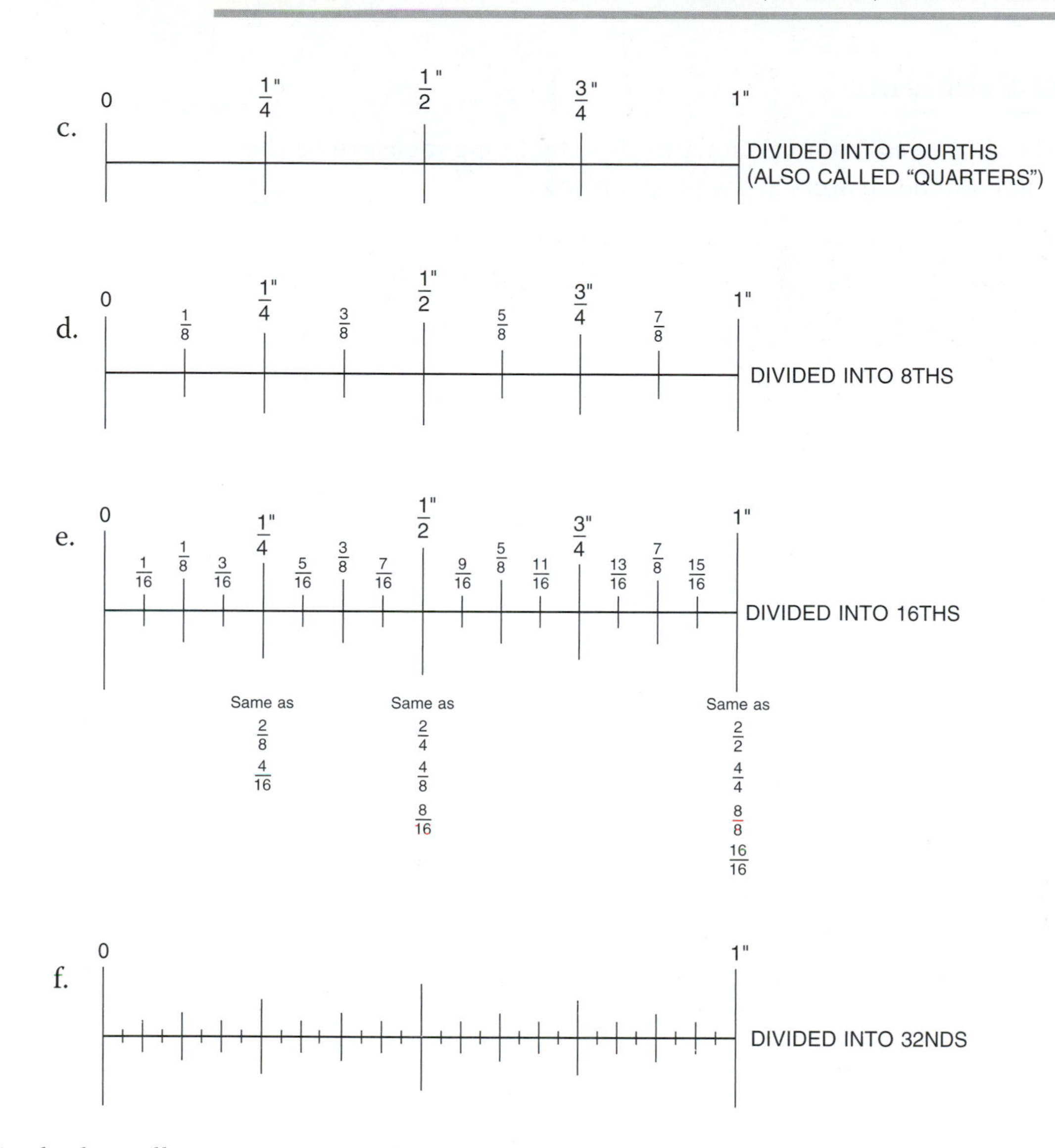

Study these illustrations: Note that the ½″ line is the longest of the marks; the ¼″ line is a little shorter; the ⅛″ line shorter still, and so on down to the smallest mark.

If the lines are not marked, as in illustration f, count the number of parts in the inch to determine the fraction of measurement.

Practical Problems

1. Read the distances from the start of this steel tape measure to the letters. Record the answers in the proper blanks.

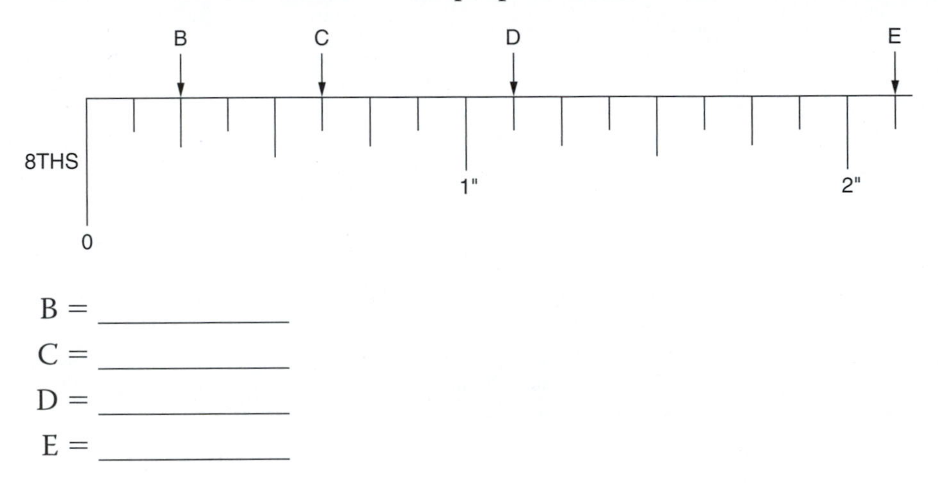

B = ____________
C = ____________
D = ____________
E = ____________

2. Read the distances from the start of this steel tape measure to the letters. Record the answers in the proper blanks.

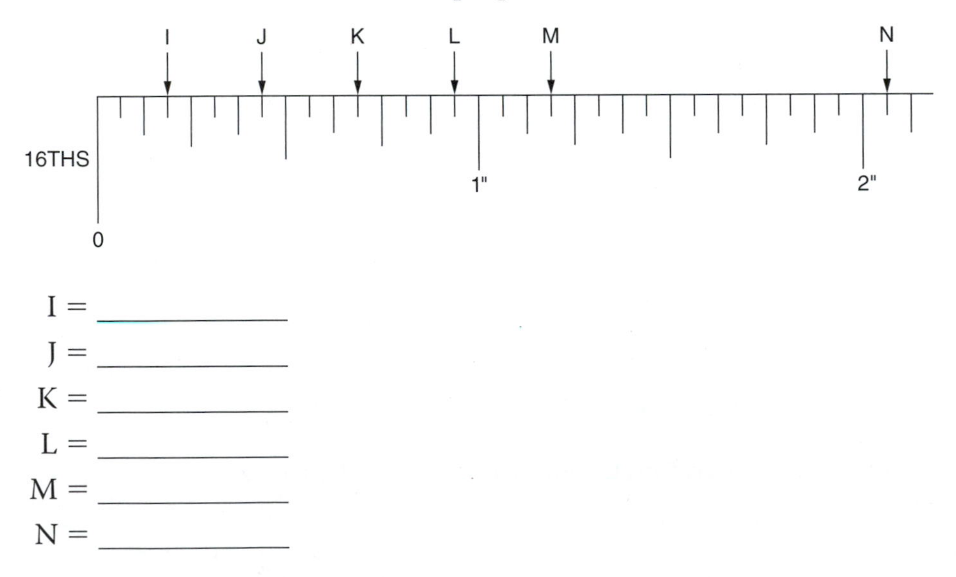

I = ____________
J = ____________
K = ____________
L = ____________
M = ____________
N = ____________

3. What is the measurement to A and B from 0″?

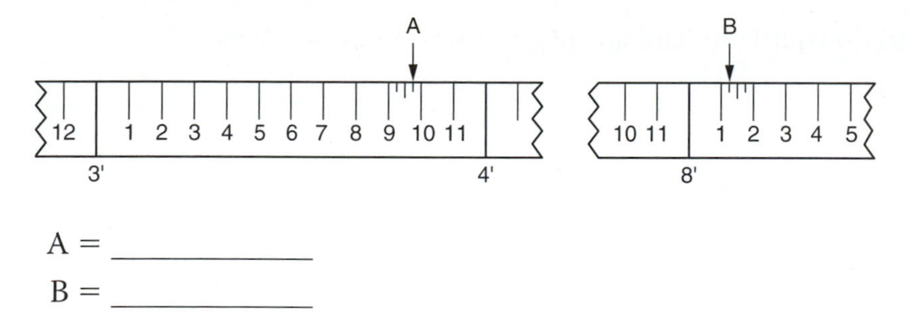

A = ____________

B = ____________

4. Using a ruler, draw lines of these lengths and label your drawings with the correct measurements.

a. $2\frac{1}{8}''$

b. $\frac{7}{16}''$

c. $1\frac{3}{4}''$

d. $3\frac{3}{16}''$

e. $2\frac{7}{8}''$

5. Draw a square (□) with sides measuring $2\frac{3}{16}''$.

6. Draw a rectangle (▭) with a length of $4\frac{7}{8}''$ and a width of $2\frac{1}{4}''$.

Classroom Exercises

Measure the following and do your work and any illustrations on these pages.

Your work table:

Length: ______________

Width: ______________

Blackboard:

Length with frame: ______________ Width with frame: ______________

Length w/o frame: ______________ Width w/o frame: ______________

Floor dimensions:

Wall-to-wall: ______________ × ______________

Baseboard-to-baseboard: ______________ × ______________

Choose 2 additional objects to measure:

____

____

Shop Exercises

Draw a picture of the following objects, label dimensions.

Symbols and abbreviations: i/s = inside, o/s = outside, ∅ = diameter

Pipe:

Length ______________

i/s ∅ ______________

o/s ∅ ______________

Wall thickness ______________

I-Beam:

Height (i/s) ______________

Leg (i/s) ______________

Plate thickness ______________

Flat plate:		**Channel:**	
Length	____________	Length	____________
Width	____________	Leg (i/s)	____________
Thickness	____________	Leg (o/s)	____________
Flange diameter *(If available in shop)*	____________	Plate thickness	____________

Choose two additional objects to measure:

____

____

Measure the following, to the best of your ability, using the tape or rule. Reminder: *Kerf* is the material cut by a saw blade, cutting torch, or other cutting tool. *Kerf width* is its measurement.

Kerf width:	**Remeasure using the Micrometer/caliper:**
Cutting torch #1 ____________	____________
Cutting torch #2 ____________	____________
Band saw ____________	____________
Blade ____________	____________
Chalk line #1 ____________	____________
Chalk line #2 ____________	____________
Drill bit ____________	____________
Tig wire ____________	____________

Micrometers and Calipers

These devices make precise measurements of items that may be difficult to read with a tape or rule. Precise measurements, accurate to 3 or 4 decimal places, are important to the machinist and engineer. The machinist has tools and cutting equipment for such precise work. Depending on the task at hand, these measurements can be helpful to the welder; however, they are normally rounded off to tenths or hundredths.

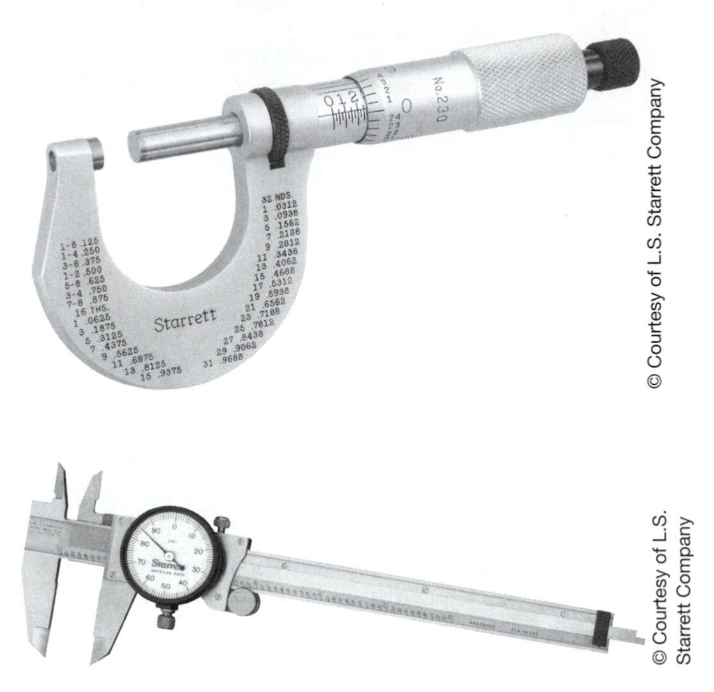

© Courtesy of L.S. Starrett Company

© Courtesy of L.S. Starrett Company

Examples of measurements taken with micrometers and calipers:

Micrometer

o/s diameter piping and tubing, wire thickness, square or rectangular tube-wall thickness, plate thickness, etc.

NOTE: Wall thickness of round pipe and tubing should not be read with a standard micrometer. Can you figure out why that is?

Caliper

i/s and o/s diameter of round pipe and tubing, i/s and o/s dimensions of square or rectangular pipe and tubing, pipe-wall thickness, plate thickness, wire dimension, etc.

NOTE: In measuring pipe or tubing wall, i/s diameters, or plate, determine if the edge of the material is flared or burred. This will distort measurements. Remove flaring and burrs.

If caliper does not have a read-out display, place the measure on a tape or rule for accuracy.

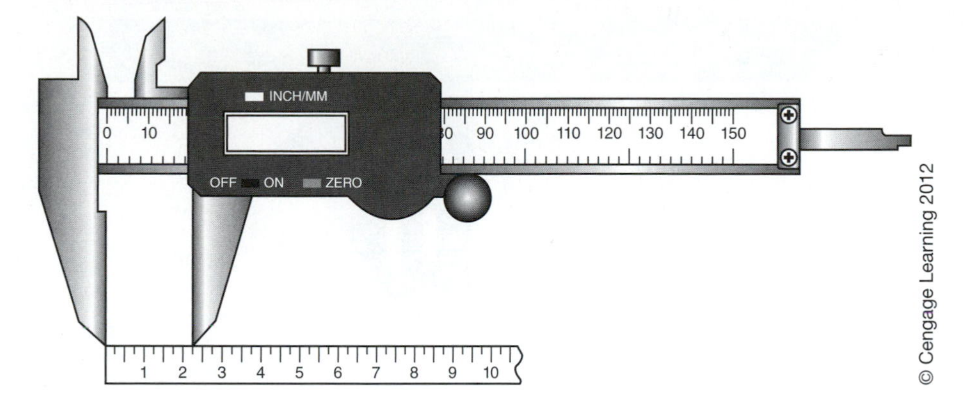

Exercises

Re-measure the previous small items with calipers or micrometer.

Additional Tools for the Welder

Square

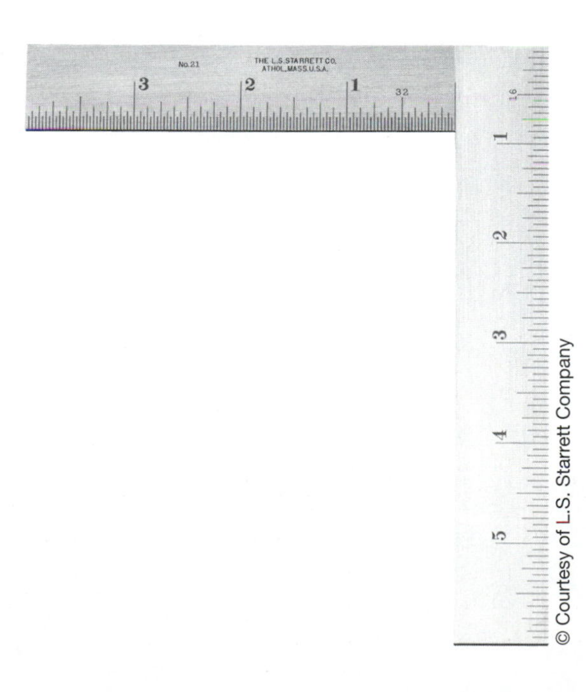

Combination square

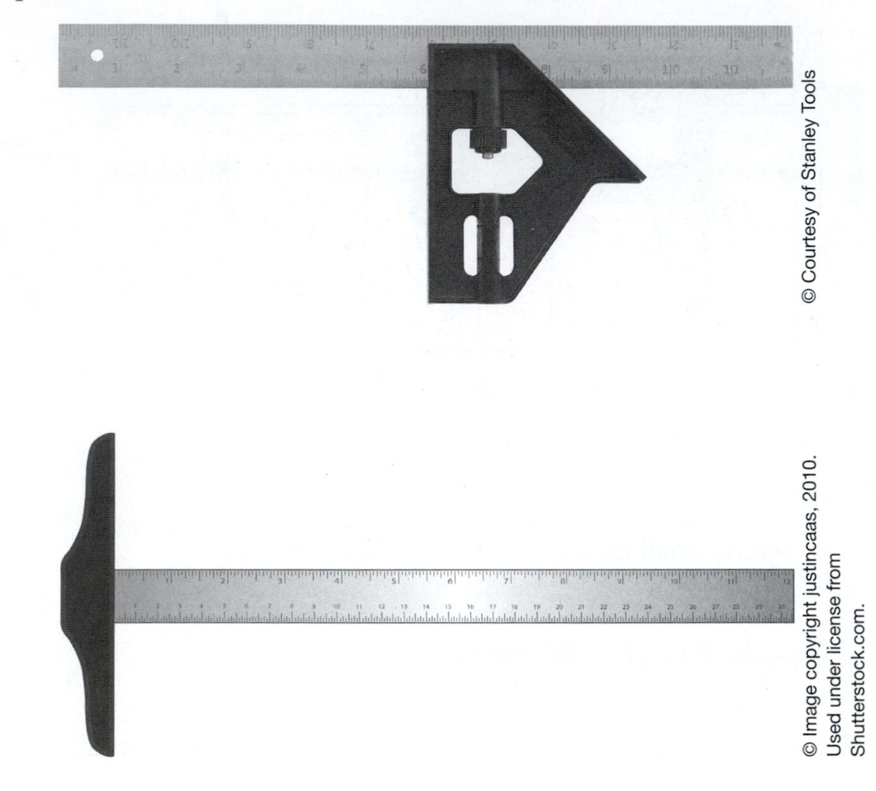

T-square

Additional tools include the plumb bob and compass.

UNIT 7

Addition of Common Fractions

Basic Principles

In order to figure dimensions or to find out how much material is needed for a particular job, the addition of common fractions is necessary. Addition, for which the symbol "plus" (+) is used, is the process of finding the total of two or more numbers or fractional parts of numbers. Fractions cannot be added if their denominators are unlike ($\frac{1}{8} + \frac{3}{4}$). Therefore, it is necessary to change all the denominators to the same quantity. This change to the bottom number can only be done with multiplication. At times, only 1 fraction needs to be changed (made larger), and at other times all need to be changed so that the bottoms are the same (common).

When adding or subtracting fractions, the least common denominator or LCD must be found. **In order to add or subtract fractions, the bottom numbers must be the same (common).**

RULE: If the bottom number of a fraction is multiplied by a number, you must also multiply the top of that fraction by the same number.

EXAMPLE: The product of the addition: $\frac{1}{8} + \frac{3}{4} =$

	Step 1		Step 2	Step 3
$\frac{1}{8}$	$\rightarrow$		$\rightarrow$	$\frac{1}{8}$
$+\frac{3}{4}$	$\frac{}{4} \times 2 = \frac{}{8}$	$\rightarrow$	$\frac{3 \times 2 = 6}{4 \times 2 = 8}$	$+\frac{6}{8}$
				$\frac{7}{8}$

STEP 1 Concentrate only on the denominator. Determine whether or not the smallest denominator, 4, can be changed into the bigger denominator, 8. In the example, the bottom number can be multiplied by 2 to change it into 8.

STEP 2 Since the bottom is multiplied by 2, we must also multiply the top by 2. $\frac{3}{4}$ is changed to $\frac{6}{8}$.

STEP 3 Add the tops together. The denominator remains the same once the common denominator 8 is found.

ANSWER: $\frac{1}{8} + \frac{3}{4} = \frac{7}{8}$

EXAMPLE 1: Add $\frac{1}{16}''$ + $\frac{1}{8}''$ + $\frac{1}{4}''$ (Remember Step 1: Concentrate only on bottom numbers at first. Then change top numbers.)

	Step 1	Step 2	Step 3	Answer
$\frac{5}{16} =$	$\frac{}{16}$ →	$\frac{}{16}$ →	$\frac{5}{16}$	$\frac{5}{16}$
$\frac{3}{8} =$	$\frac{}{8} \times 2 = \frac{}{16}$ →	$\frac{3 \times 2}{8 \times 2} = \frac{6}{16}$ →	$\frac{6}{16}$	$\frac{6}{16}$
$+\frac{1}{4} =$	$\frac{}{4} \times 4 = \frac{}{16}$ →	$\frac{1 \times 4}{4 \times 4} = \frac{4}{16}$ →	$+\frac{4}{16}$	$+\frac{4}{16}$
				$\frac{15}{16}$

EXAMPLE 2: In this example, both fractions are changed by multiplying the denominators.

$\frac{2}{5} + \frac{1}{3} =$

$\frac{2}{5}$ →	$\frac{}{5 \times 3}$ →	$\frac{}{15}$ →	$\frac{2 \times 3}{15}$ →	$\frac{6}{15}$	$\frac{6}{15}$
$+\frac{1}{3}$	$\frac{}{3 \times 5}$	$\frac{}{15}$	$\frac{1 \times 5}{15}$	$\frac{5}{15}$	$+\frac{5}{15}$
					$\frac{11}{15}$

ANSWER: $\frac{2}{5} + \frac{1}{3} = \frac{11}{15}$

To add mixed numbers, add the whole numbers and fractions separately, then continue the sums.

PROBLEM:

$$2\frac{2}{3} + 4\frac{1}{2} + 7\frac{1}{6}$$

Add the whole numbers.

$$\begin{array}{r} 2\frac{2}{3} \\ 4\frac{1}{2} \\ +7\frac{1}{6} \\ \hline 13 \end{array}$$

Add the fractions after finding the LCD.

$$\begin{array}{cccc} \frac{2}{3} & \overline{3}\times 2 = \overline{6} & \frac{2 \times 2}{3 \times 2} = \frac{4}{6} & \frac{4}{6} \\ \frac{1}{2} & \overline{2}\times 3 = \overline{6} & \frac{1 \times 3}{2 \times 3} = \frac{3}{6} & \frac{3}{6} \\ +\frac{1}{6} & \rightarrow & \frac{1}{6} & +\frac{1}{6} \\ \hline & & & \frac{8}{6} \end{array}$$

Reduce the improper fraction to its lowest terms by dividing the denominator into the numerator. Reduce the final fraction if possible.

$$\frac{8}{6} \quad 6\overline{)8} \quad \begin{array}{r} 1 \\ 6\overline{)8} \\ -6 \\ \hline 2 \end{array} = 1\frac{2}{6} = 1\frac{1}{3}$$

Add the mixed number that results to the sum of the whole numbers.

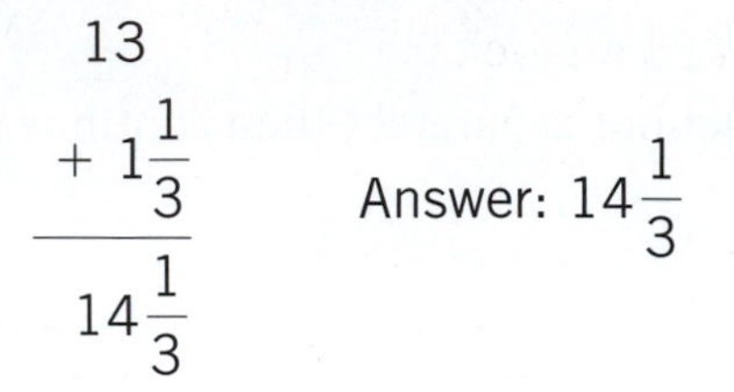

$$\begin{array}{r} 13 \\ +\,1\frac{1}{3} \\ \hline 14\frac{1}{3} \end{array}$$

Answer: $14\frac{1}{3}$

Exercises

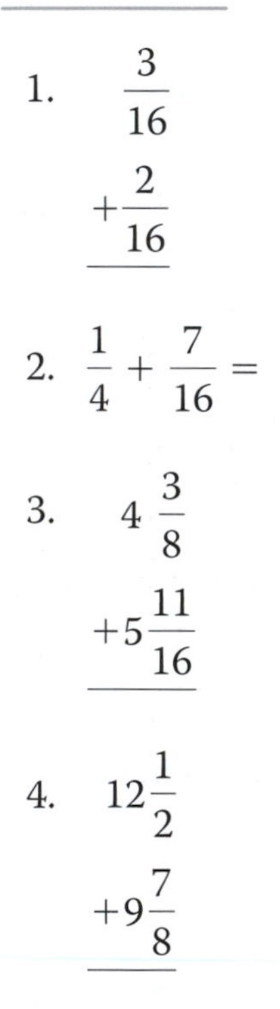

1. $\begin{array}{r} \frac{3}{16} \\ +\frac{2}{16} \\ \hline \end{array}$ ____________

2. $\frac{1}{4} + \frac{7}{16} =$ ____________

3. $\begin{array}{r} 4\frac{3}{8} \\ +5\frac{11}{16} \\ \hline \end{array}$ ____________

4. $\begin{array}{r} 12\frac{1}{2} \\ +9\frac{7}{8} \\ \hline \end{array}$ ____________

Practical Problems

1. Find the total combined length of these 2 pieces of bar stock. ______________

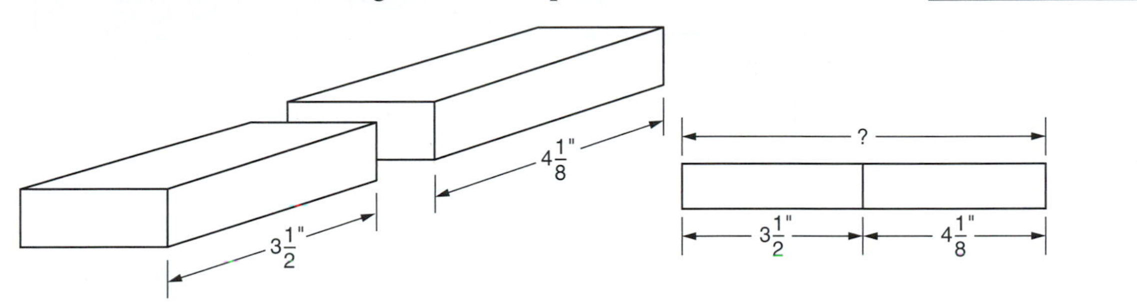

2. If you stack the 2 pieces of steel bar, what is the height of the stack? ______________

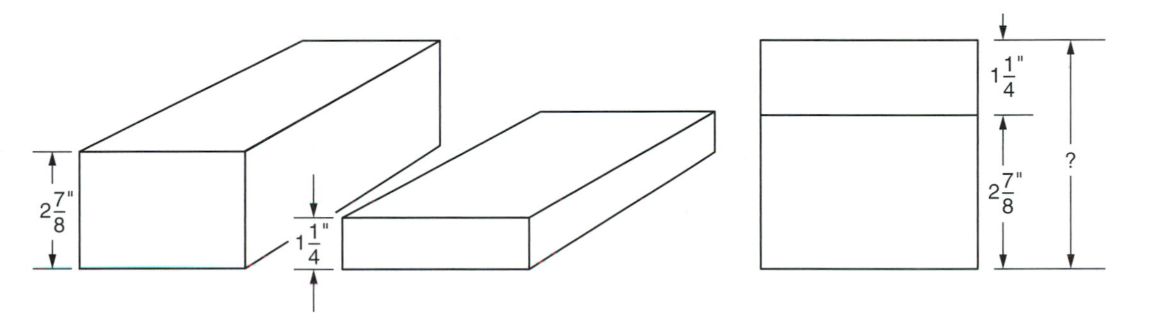

3. Find the total combined weight of these 3 pieces of steel. ______________

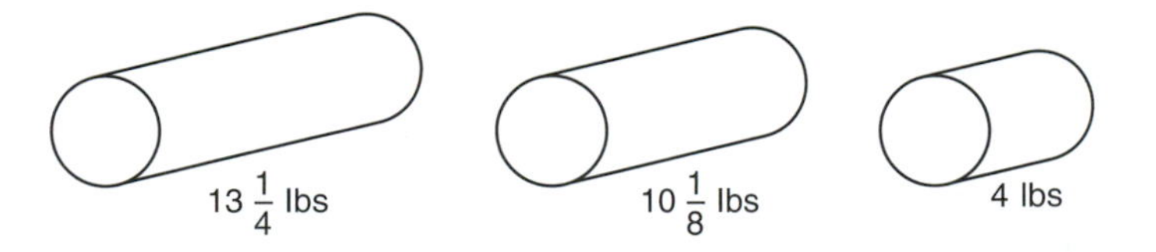

4. Four holes are drilled in this piece of flat stock. What is the total combined distance between the centers of the end holes? ________

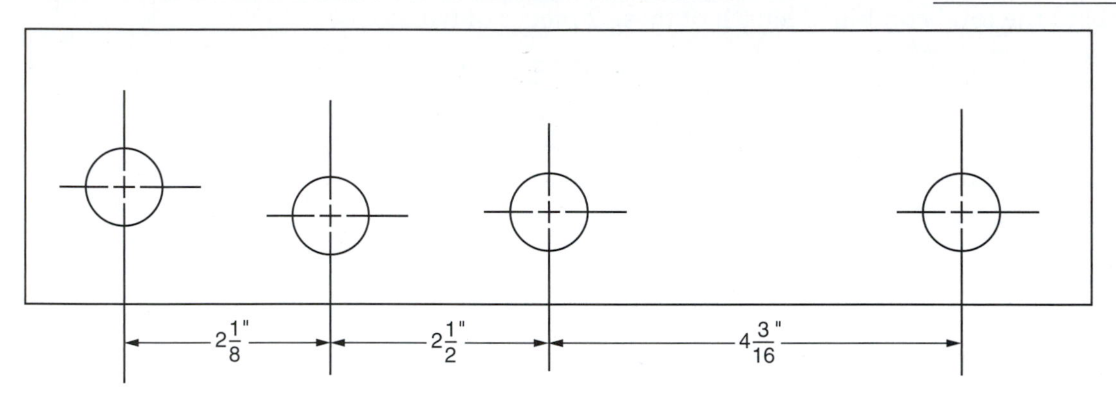

5. Two circular pieces of steel are placed side by side. What is their combined length? ________

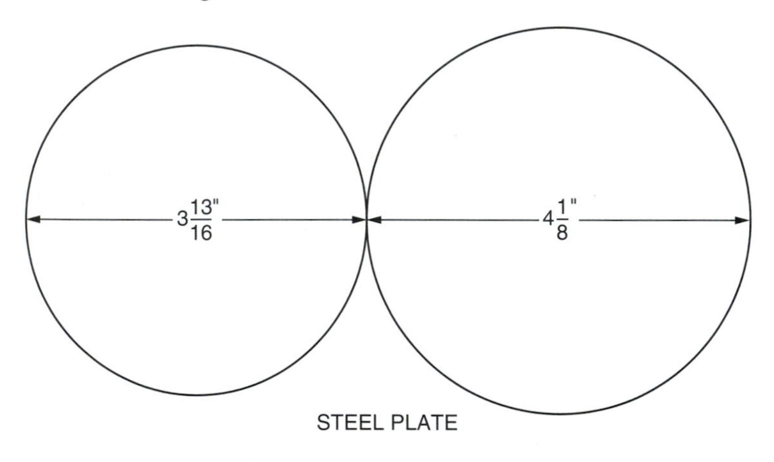

6. What is the length of this weldment? ________

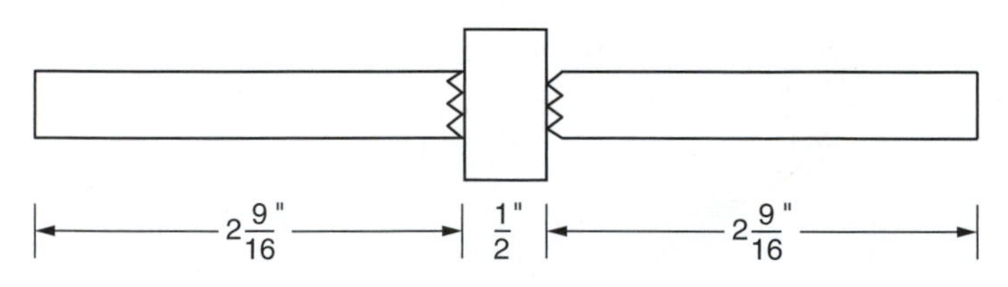

7. To make shims for leveling equipment, three pieces of material are welded together. What is the total thickness of the welded material, in inches? ______

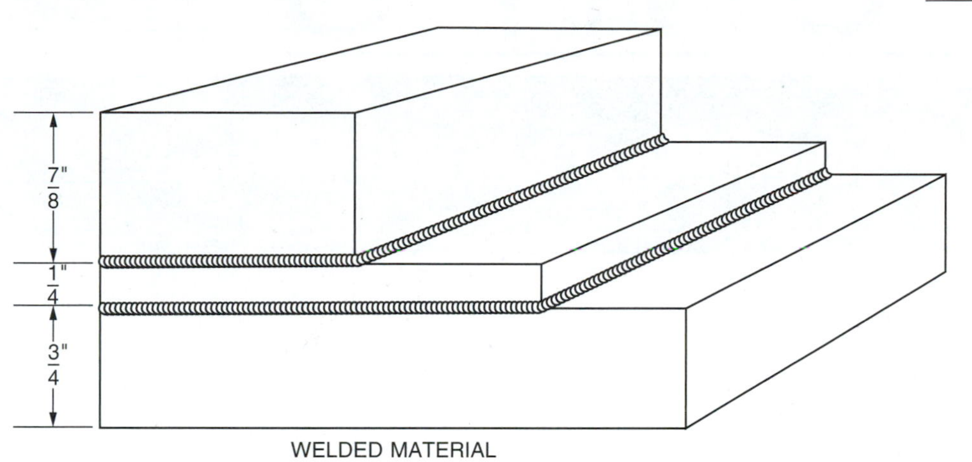

WELDED MATERIAL

8. To make use of some scrap, four pieces of 1½″ cold-rolled bar are welded together. What is the total length of the completed weldment? ______

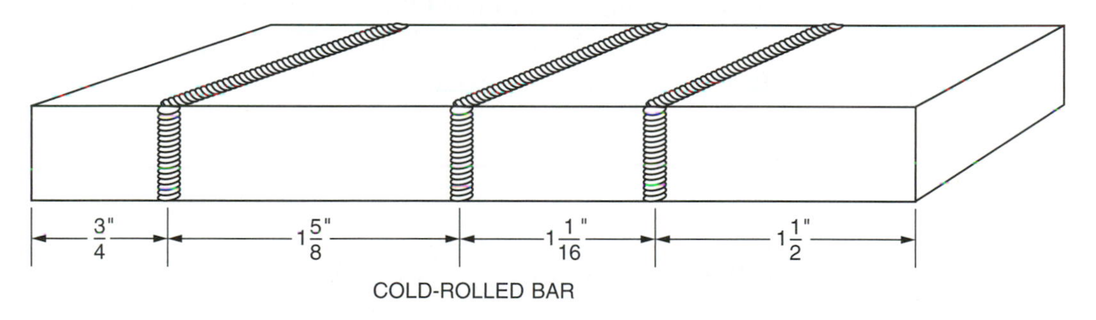

COLD-ROLLED BAR

9. What is the length of the weldment if pieces A and B are welded together? ______

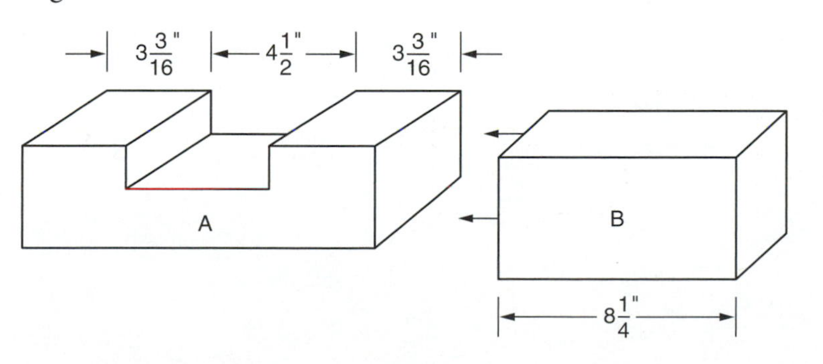

UNIT 8

Subtraction of Common Fractions

Basic Principles

Sometimes measurements not on the blueprint are needed, and it may be necessary to subtract one fractional measurement from another to obtain the correct length of the materials.

RULE: The steps in subtraction of fractions are similar to addition. All fractions must have common denominators. We subtract the indicated top number from the first instead of adding them together. Borrowing may be necessary.

EXAMPLE: $\frac{7}{8} - \frac{1}{4} = ?$

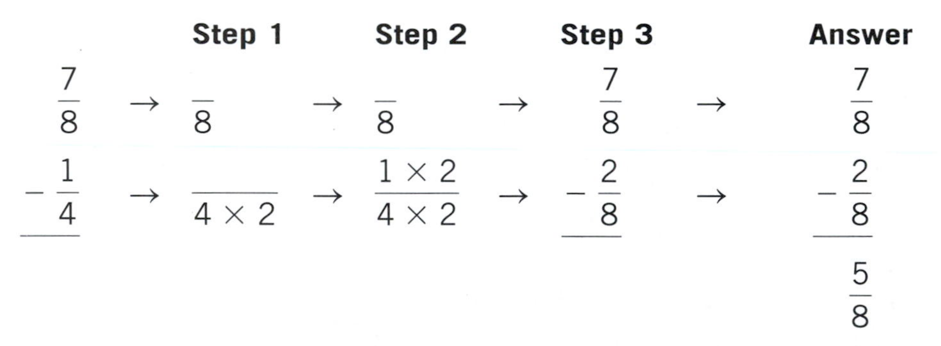

		Step 1		Step 2		Step 3		Answer
$\frac{7}{8}$	→	$\frac{\ }{8}$	→	$\frac{\ }{8}$	→	$\frac{7}{8}$	→	$\frac{7}{8}$
$-\frac{1}{4}$	→	$\frac{\ }{4 \times 2}$	→	$\frac{1 \times 2}{4 \times 2}$	→	$-\frac{2}{8}$	→	$-\frac{2}{8}$
								$\frac{5}{8}$

EXAMPLE: Subtract $1\frac{3}{4}$ from $3\frac{1}{2}$.

$$\begin{array}{r} 3\frac{1}{2} \\ -1\frac{3}{4} \\ \hline \end{array}$$

These fractions have unlike denominators. In this example, we can change ½ into fourths.

$$3\frac{1}{2} \rightarrow 3\frac{1 \times 2}{2 \times 2} \rightarrow 3\frac{2}{4}$$

$$-1\frac{3}{4} \rightarrow -1\frac{3}{4}$$

Three-fourths cannot be subtracted from two-fourths. We need more fourths. Borrow a whole number from the number 3 and convert it to fourths. 1 = 4⁄4. Add the 4⁄4 to 2⁄4. We now have a total of 6⁄4. The result is 2 6⁄4. Continue with subtraction.

$$2\not{3}\frac{2}{4} + \frac{4}{4} \rightarrow 2\frac{6}{4} \qquad 2\frac{6}{4}$$

$$-1\frac{3}{4} \rightarrow -1\frac{3}{4} \qquad -1\frac{3}{4}$$

$$1\frac{3}{4}$$

Remember, we only need to borrow 1 whole number, and the number can be converted into any fraction needed. For example, 1 = 8⁄8, 1 = 16⁄16, 1 = 3⁄3, 1 = 4⁄4, 1 = 32⁄32, and so on.

Practical Problems

1. Determine the missing dimension on this welded bracket. ______

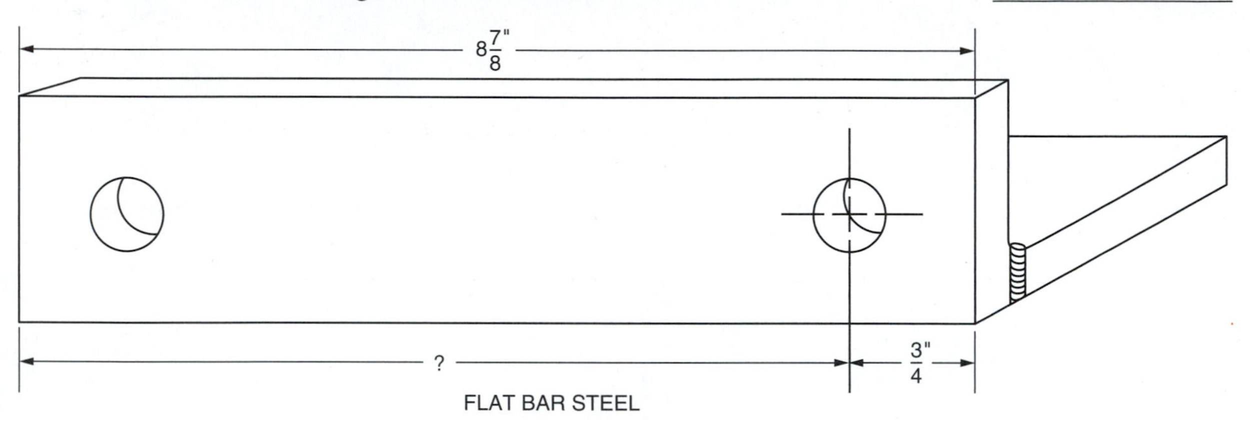

2. A $3\frac{1}{16}$" piece is cut from the steel angle iron illustrated. If there is $\frac{1}{8}$" waste caused by the kerf of the oxy-acetylene cutting process, what is the length of the remaining piece of angle iron? ______

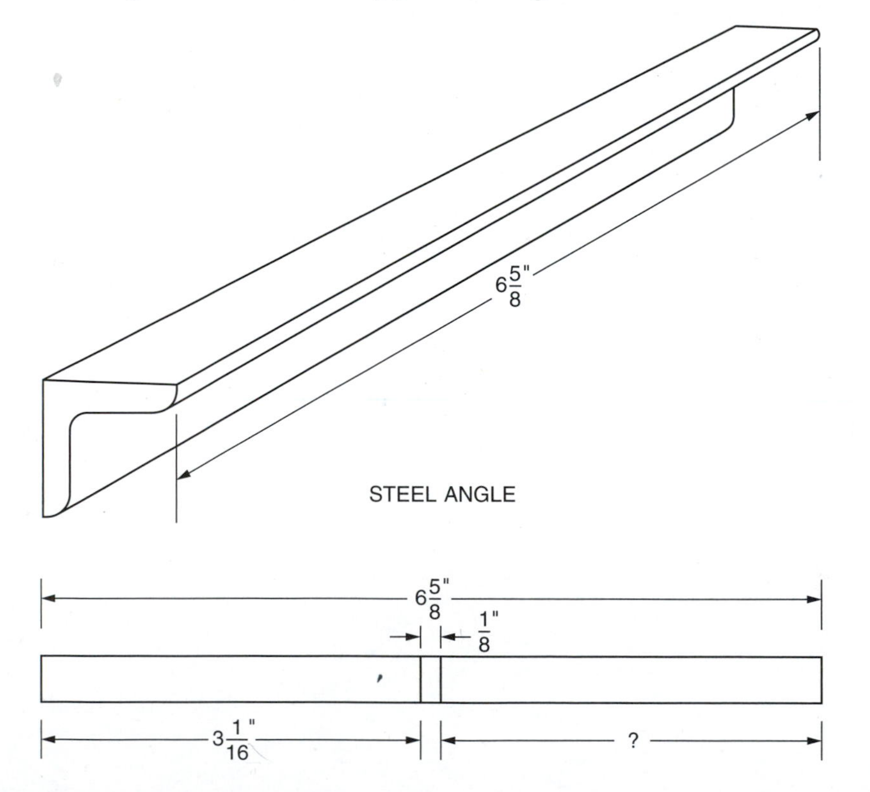

3. A $9\frac{5}{16}''$ length of bar stock is cut from this piece. What is the length of the remaining bar stock? Disregard waste caused by the width of the kerf (a). a. ____________

   Calculate remaining bar stock using $\frac{1}{4}''$ kerf width (b). b. ____________

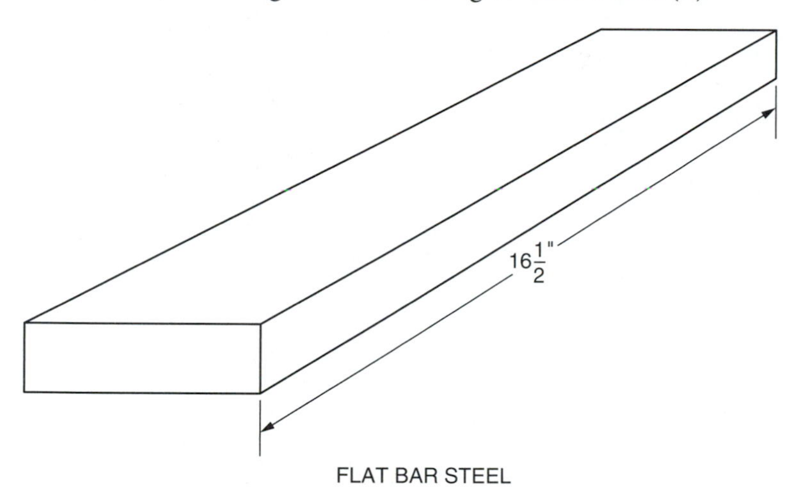

FLAT BAR STEEL

4. A $15\frac{5}{16}''$ diameter circle is flame-cut from this steel plate. Find the missing dimension. The width of the kerf is $\frac{1}{16}''$. ____________

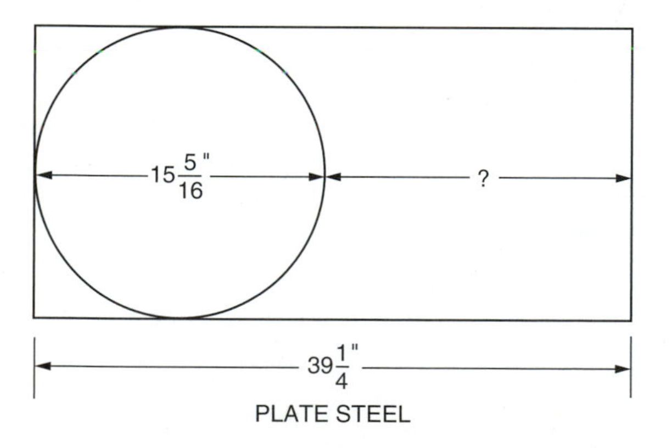

PLATE STEEL

5. Find dimension A on this steel angle. ________

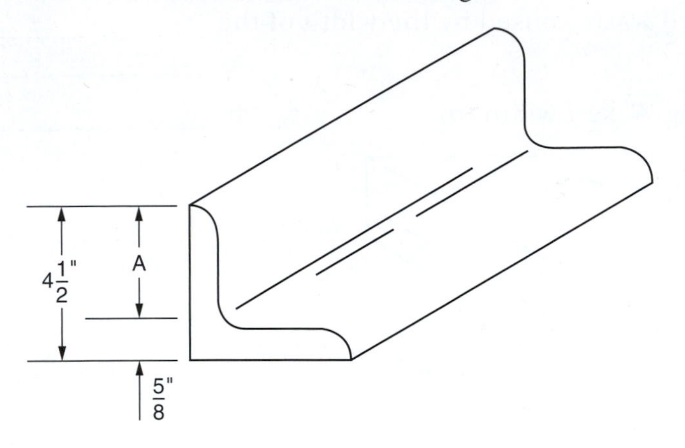

6. What is the missing dimension? ________

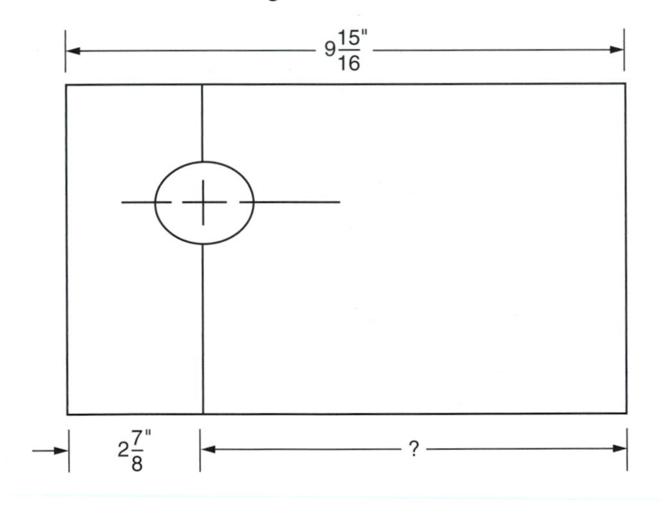

7. A flame-cut wheel is to have the shape shown. Find the missing dimension. ______

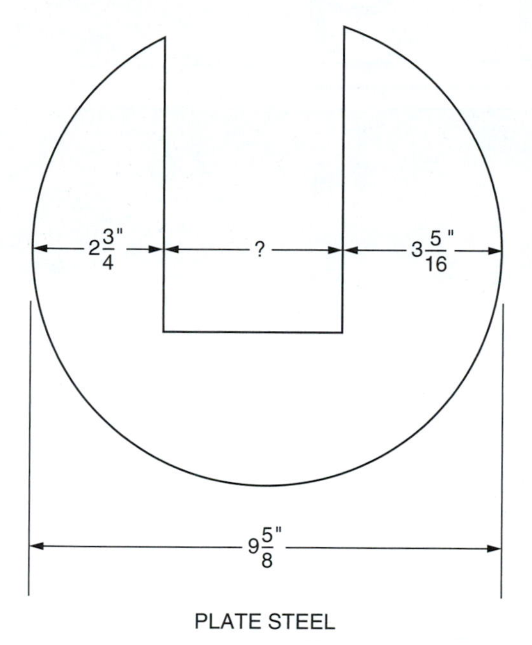

UNIT 9

Multiplication of Common Fractions

Basic Principles

The symbol for multiplication is "times" ($\times$). It is the short method of adding a number to itself a certain number of times.

Multiplication can be shown several ways:

$$7 \times 5 \qquad 7 \cdot 5 \qquad (7)5 \qquad 7(5) \qquad (7)(5)$$

The steps for multiplication and division are similar. There is only one extra step necessary in division. These steps are not the same as in addition and subtraction, so you do not need to focus on the bottom numbers of the fractions.

Multiplication

RULE: All numbers must be in fraction form.

STEP 1 Change any whole numbers or mixed numbers into improper fractions. This step is done first before moving to step 2. To change a mixed number into a fraction, multiply the bottom number times the whole number, and then add the top number. This becomes the new top number. Keep the same bottom number.

EXAMPLE: $2\frac{1}{4} \times 5\frac{2}{3}$

$2\frac{1}{4} \rightarrow$ (bottom × whole) $4 \times 2 = 8 \rightarrow$ add the top: $8 + 1 = 9$

9 will be the top number. The original bottom number is not changed.

$$2\frac{1}{4} = \frac{9}{4}$$

$5\frac{2}{3} \rightarrow$ (bottom × whole) $3 \times 5 = 15 \rightarrow$ add the top: $15 + 2 = 17$

17 will be the top number. The original bottom number is not changed.

$$5\frac{2}{3} = \frac{17}{3}$$

In original form: $2\frac{1}{4} \times 5\frac{2}{3}$

Ready for Step 2: $\frac{9}{4} \times \frac{17}{3}$

STEP 2 Cross-reduce, if possible. Reduce the top number of any fraction and the bottom number of a different fraction by dividing them by any same number that will work. This is similar to reducing fractions. Check out each top/bottom pair separately.

Check out first pair. A same number will not divide into both 4 and 17 (2 and 4 are the only possibilities).

$$\frac{9}{4} \times \frac{17}{3} \qquad \frac{}{4} \times \frac{17}{} =$$

Check out the other pair. Both 9 and 3 can be divided by 3.

$$\frac{9}{} \times \frac{}{3} \rightarrow \frac{9 \div 3}{} \times \frac{}{3 \div 3} \rightarrow \frac{\overset{3}{\cancel{9 \div 3}}}{} \times \frac{}{\underset{1}{\cancel{3 \div 3}}}$$

Rewrite: $\frac{\overset{3}{\cancel{9 \div 3}}}{4} \times \frac{17}{\underset{1}{\cancel{3 \div 3}}} \rightarrow \frac{3}{4} \times \frac{17}{1}$

STEP 3 Multiply top number × top number. This becomes the top number of the new fraction.

$$\frac{\overset{3}{\cancel{9 \div 3}}}{} \times \frac{17}{} = \frac{51}{}$$

Multiply bottom number × bottom number and place the result at the bottom of the new fraction.

$$\frac{}{4} \times \frac{}{\underset{1}{\cancel{3 \div 3}}} = \frac{}{4}$$

$$\frac{\overset{3}{\cancel{9 \div 3}}}{4} \times \frac{17}{\underset{1}{\cancel{3 \div 3}}} = \frac{51}{4}$$

Change improper fraction answer to proper form (divide bottom into top).

$$\frac{51}{4} \rightarrow 4\overline{)51} \rightarrow \begin{array}{r} 12 \\ 4\overline{)51} \\ \underline{-4} \\ 11 \\ \underline{-8} \\ 3 \end{array} \rightarrow \begin{array}{r} 12 \\ 4\overline{)51} \\ \underline{-4} \\ 11 \\ \underline{-8} \\ 3 \end{array} = 12\frac{3}{4}$$

ANSWER: $2\frac{1}{4} \times 5\frac{2}{3} = 12\frac{3}{4}$

Exercises

1. $3\frac{1}{2} \times 8\frac{3}{8}$ __________

2. $7\frac{1}{16} \times 6$ __________

Practical Problems

1. If 5 pieces of steel bar each 6½″ long are welded together, how long will the new bar be? __________

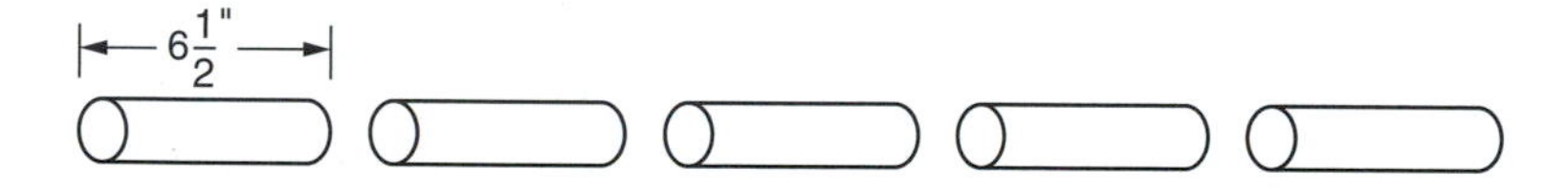

2. A welder has an order for 8 pieces of angle iron, each 7¼″ long. What is the total length of the angle needed to complete the order? Disregard waste per cut. __________

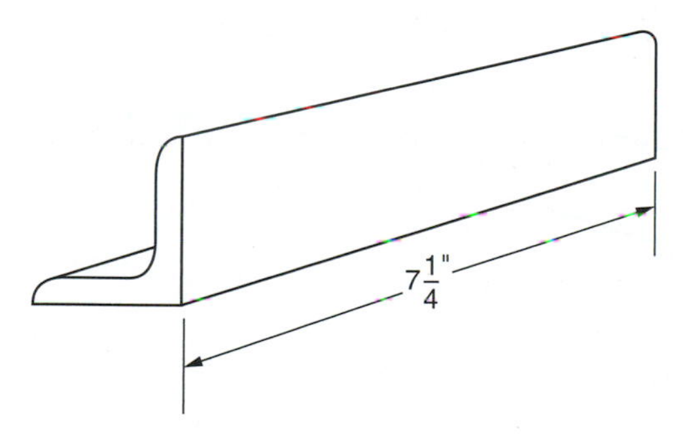

STEEL ANGLE

3. Three of these welded brackets are needed. What is the total length, in inches, of the bar stock needed for all of the brackets? __________

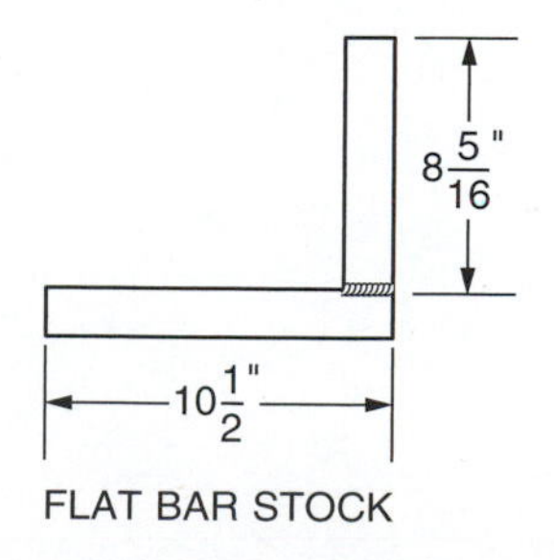

4. Twenty-two pieces, each 6½″ long, are cut from this steel angle. There is ⅛″ kerf on each cut. How much angle remains after the 22 pieces are cut? ________

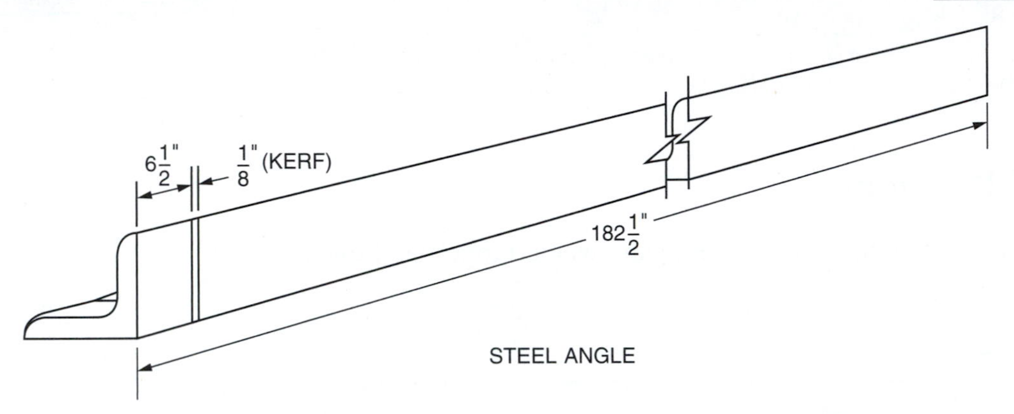

STEEL ANGLE

5. Seven pieces of ½″ round stock, each 3″ long, are cut from a bar. How much material is required? Allow ⅛″ waste for each cut. ________

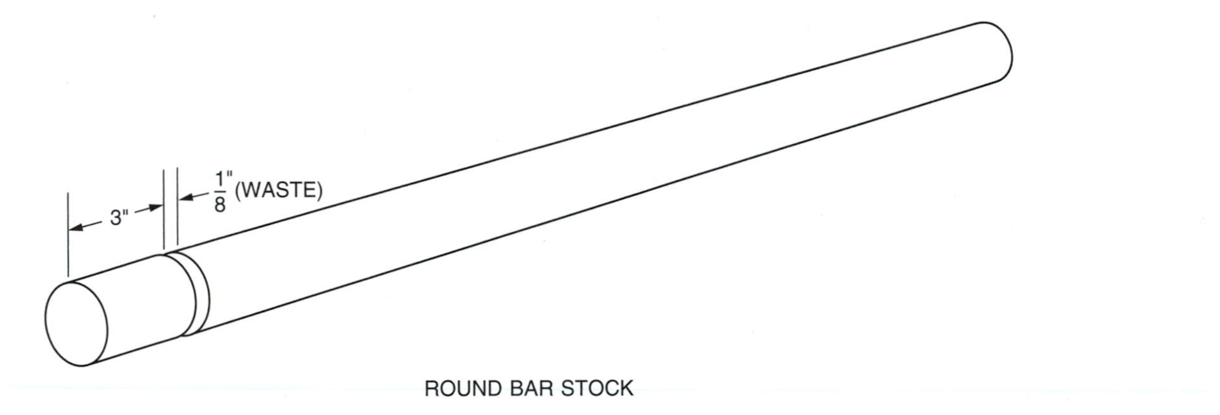

ROUND BAR STOCK

6. Thirteen pieces of steel angle, each $6\frac{7}{8}''$ long, are welded to a piece of flat bar for use as concrete reinforcement. What is the total length of steel angle required? Allow $\frac{3}{16}''$ waste per cut. ____________

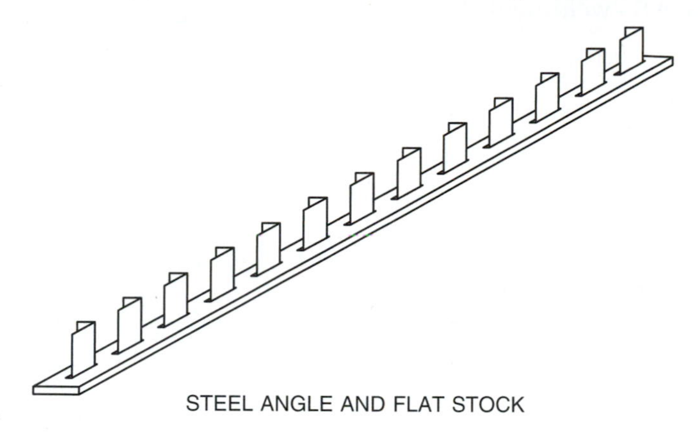

STEEL ANGLE AND FLAT STOCK

7. To weld around this weldment, $16\frac{1}{2}$ arc rods are needed. If $6\frac{3}{4}$ of the weldments are completed in an 8-hour shift, how many arc rods will be needed? ____________

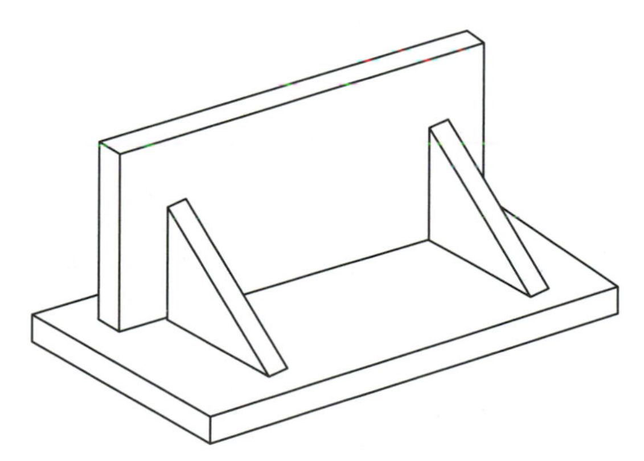

8. It takes $5\frac{3}{4}$ rods to weld the upright to the base plate. How many rods are needed to make 17 weldments? ____________

   How many rods are needed to make 85 weldments? ____________

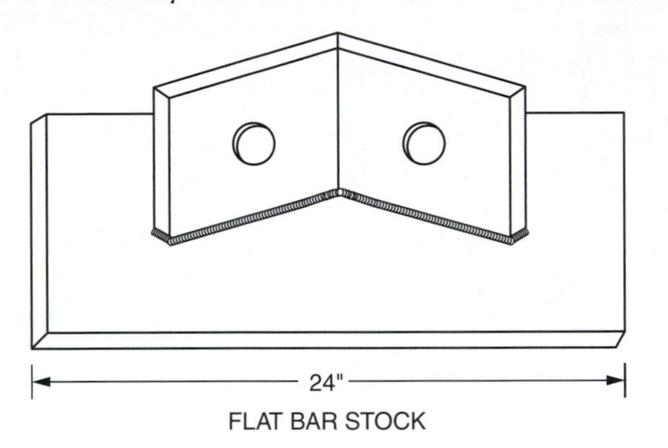

UNIT 10

Division of Common Fractions

Basic Principles

Division has the same steps as in multiplication of fractions, except Step 1-a as shown, which changes division back into multiplication.

RULE: Invert the divisor, then multiply. (Invert means "turn upside down." For example, ¾ inverted is ⁴⁄₃.)

EXAMPLE: $15\frac{1}{2} \div 2\frac{3}{4}$

STEP 1 Same as in multiplication, all numbers must be in fraction form. Change any whole or mixed numbers into fractions. This must be done first before you go to Step 1-a.

$15\frac{1}{2} \div 2\frac{3}{4}$

$2 \times 15 = 30 \qquad 30 + 1 = 31 \qquad 15\frac{1}{2} = \frac{31}{2}$

$4 \times 2 = 8 \qquad 8 + 3 = 11 \qquad 2\frac{3}{4} = \frac{11}{4}$

Thus, $\frac{31}{2} \div \frac{11}{4}$

STEP 1a Change the division sign into multiplication (×) and invert (flip) the following fraction. Do not flip the fraction in front of the sign. You are now back into multiplication. Follow multiplication rules.

$$\frac{31}{2} \div \frac{11}{4} = \frac{31}{2} \times \mathbf{\frac{4}{11}}$$

STEP 2 Cross-reduce, if possible.

$$\frac{31}{2} \times \frac{4}{11} \rightarrow \frac{31}{2 \div 2} \times \frac{4 \div 2}{11} \rightarrow \frac{31}{1} \times \frac{2}{11}$$

STEP 3 Multiply tops together. This becomes the top of the answer. Multiply bottoms together. This becomes the bottom of the answer.

$$\frac{31}{1} \times \frac{2}{11} = \frac{62}{11}$$

 Reduce as needed. Put the answer in proper form.

$$\frac{62}{11} \rightarrow 11\overline{)62} \rightarrow \begin{array}{r} 5 \\ 11\overline{)62} \\ 55 \end{array} \rightarrow \begin{array}{r} 5\frac{7}{11} \\ 11\overline{)62} \\ \underline{55} \\ 7 \end{array} \quad \frac{7}{11} \text{ cannot be reduced.}$$

$$\frac{62}{11} = 5\frac{7}{11}$$

Exercises

1. $\frac{3}{4} \div \frac{5}{8}$ ________________

2. $5\frac{1}{3} \div 2$ ________________

3. $38\frac{1}{2} \div 12\frac{1}{3}$ ________________

Practical Problems

1. A 36″ piece of steel angle is in stock. How many 5½″ pieces can be cut from it? (36 ÷ 5½). Disregard width of cut (a). Recalculate with a ¼″ width of cut (b).

   a. ______________

   b. ______________

2. It is necessary to cut as many keys as possible to fit this keyway. A piece of key stock, 12⅛″ long, is in stock. How many full-sized keys can be sheared from it? ______________

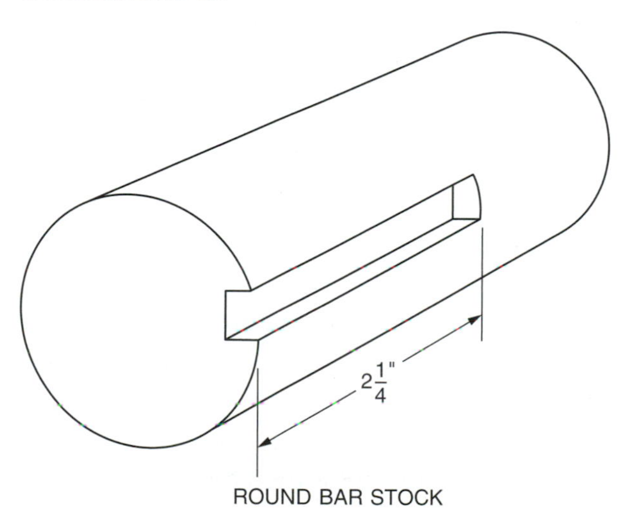

3. This piece of angle is to be used for an anchor bracket. If the holes are equally spaced, what is the measurement between hole 1 and hole 2? ______________

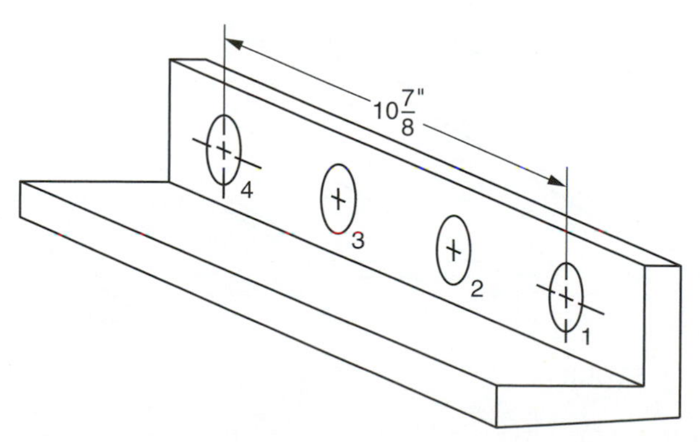

4. Each pre-drilled bar of angle iron in problem 3 weighs 17¾ pounds. If 284 pounds of angle iron are in the stock pile, how many bars are in stock? __________

5. A piece of 16-gauge sheet metal 24¼″ wide is in stock.

   a. How many ¾″ strips can be sheared from this sheet? a. __________

   b. What size piece is left over? b. __________

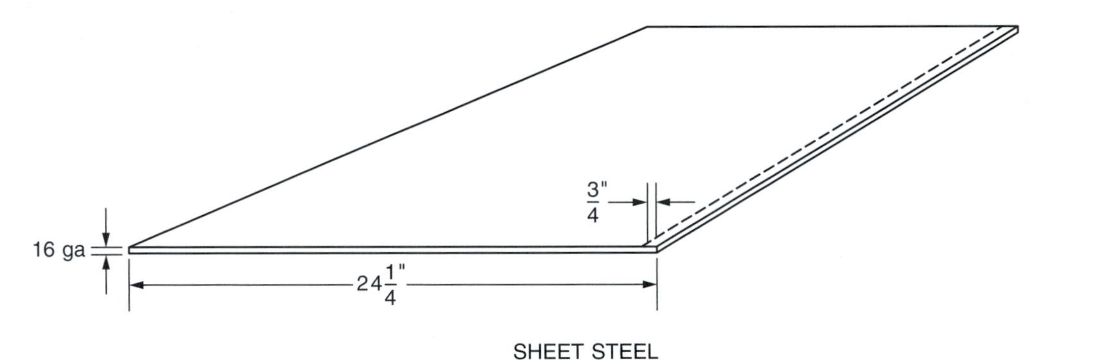

SHEET STEEL

6. How many 2¼″ long pieces can be cut from a 14¾″ length of channel iron? Kerf width is ⅛″. __________

7. Three bars of steel are shown. How many pieces, each 21½″ long, can be cut from the total length of the three bars after they are joined by welding? Disregard width of the cut. __________

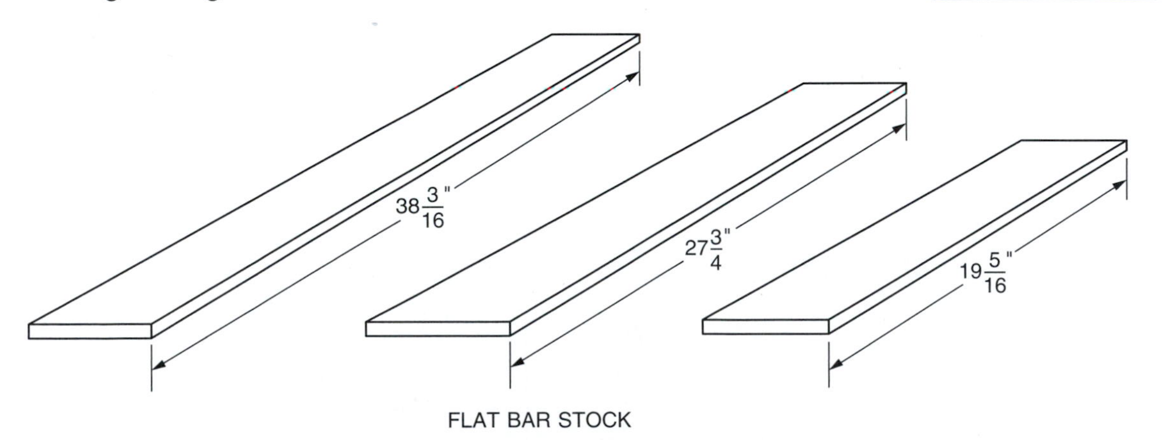

FLAT BAR STOCK

UNIT 11

Combined Operations with Common Fractions

Basic Principles

Combined operations include addition, subtraction, multiplication, and/or division. Apply these to solve the following problems.

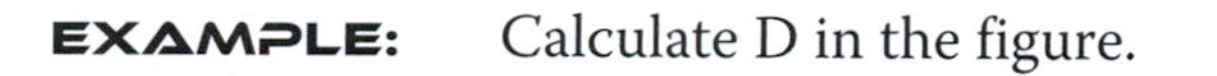

EXAMPLE: Calculate D in the figure.

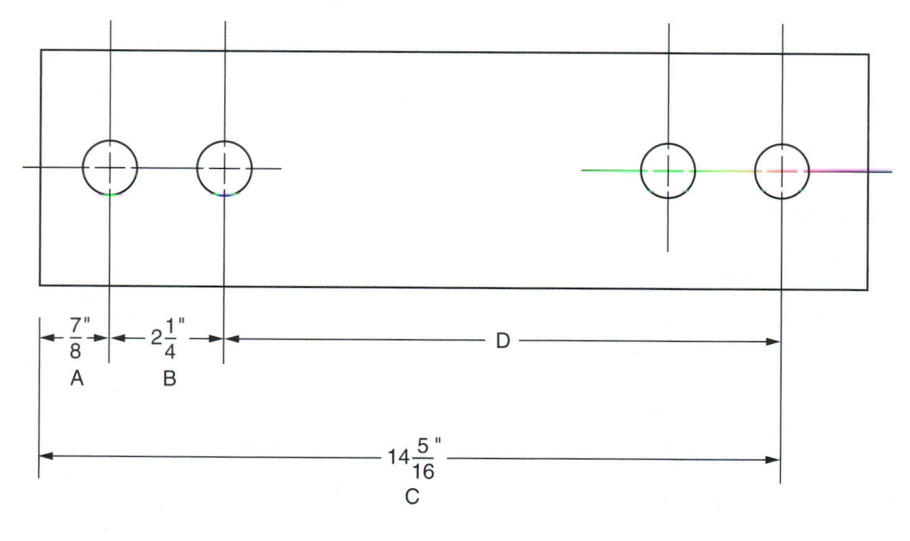

To solve:
First, add A and B.

$$\begin{array}{r} \frac{7}{8} \\ +2\frac{1}{4} \\ \hline \end{array} \rightarrow \begin{array}{r} \frac{7}{8} \\ +2\frac{1\times2}{4\times2} \\ \hline \end{array} \rightarrow \begin{array}{r} \frac{7}{8} \\ +2\frac{2}{8} \\ \hline 2\frac{9}{8} \end{array} \quad 2\frac{9}{8} = 3\frac{1}{8}''$$

Next, subtract 3⅛″ from C.

$$\begin{array}{r} 14\frac{5}{16} \\ -\ 3\frac{1}{8} \\ \hline \end{array} \rightarrow \begin{array}{r} 14\frac{5}{16} \\ -3\frac{1\times2}{8\times2} \\ \hline \end{array} \rightarrow \begin{array}{r} 14\frac{5}{16} \\ -3\frac{2}{16} \\ \hline \end{array} = 11\frac{3}{16}$$

ANSWER: $D = 11\frac{3}{16}''$

Practical Problems

1. Nine sections of steel bar, each 13¼″ long, are welded together. The finished piece is cut into 4 equal parts. What is the length of each new piece? Disregard cut waste. ________

2. Divide a 38⅛″ piece of mild steel into 6 equal parts. What is the length of 3 parts when welded together? Disregard cut waste. ________

3. Cut 3 pieces, each 14⅛″ long, from a pipe 50¾″ long. The kerf of the cut is ⅛″.

   a. What is the combined length of the 3 pieces? a. ______________

   b. What is the length of the waste? b. ______________

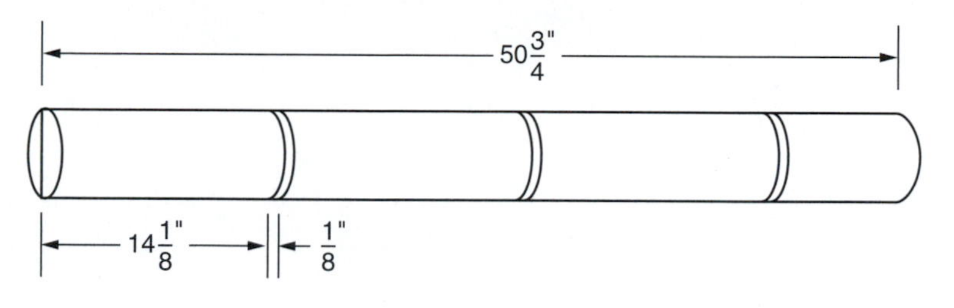

SECTION 3

DECIMAL FRACTIONS

UNIT 12

Introduction to Decimal Fractions and Rounding Numbers

Basic Principles

Decimal fractions are similar to common fractions in that they describe part of a whole object.

In decimals, an object is divided into tenths, hundredths, thousandths, etc. Welders, however, primarily work with tenths and hundredths, mandated by the accuracy of the tools of the trade.

NOTE: *For all decimal problems in this workbook, round to hundredths (two places unless otherwise noted. You may round to three or four places if that place number is a 5 (i.e., .125 or .0625). Greater accuracy is achieved only if the final answer is rounded off, not the numbers used to arrive at the answer.*

A decimal point separates the whole numbers from the parts of a whole. Whole numbers are always to the left of the decimal point.

The first place after the decimal point is called tenths, the second place is called hundredths, the third place is called thousandths.

EXAMPLE: .758

Tenths	Hundredths	Thousandths
.7	5	8

Tenths describes 1 whole object divided into 10 equal parts.

Hundredths describes 1 whole object divided into 100 equal parts.

Examples of tenths:

.3 .7 .9

Example of hundredths:

.36 .16 .04

Rounding Off Decimals

Rounding off helps express measurements according to the needs of our trade. Welders generally round off to the nearest tenths or hundredths, or whole numbers.

RULE:

Rule in rounding to tenths:

If the second place number is 5 or greater, increase the tenth by 1. If the second place number is 4 or less, the tenth does not change.

EXAMPLES:

.68 rounded to tenths is .7

.64 rounded to tenths is .6

Rule in rounding to hundredths:

If the third place number is 5 or greater, increase the hundredth by 1. If the third place number is 4 or lower, the hundredth does not change.

EXAMPLES:

.357 rounded to hundredths is .36

.351 rounded to hundredths is .35

Rule in rounding to whole numbers:

If the first place number is 5 or greater, increase the whole number by 1. If the first place number is 4 or lower, the whole number stays the same.

EXAMPLES:

$123.61 = 124 \qquad 18.39 = 18$

RULE: Zeroes placed at the end of a decimal have no effect on the value.

EXAMPLES:

$.5 = .50 \qquad .50 = .500$

Zeroes placed in front of the decimal point have no effect on the value as long as there are no whole numbers.

EXAMPLE: $.25 = 0.25$

Practical Problems

1. A mile is divided into tenths. Express each distance as a decimal fraction of a mile.

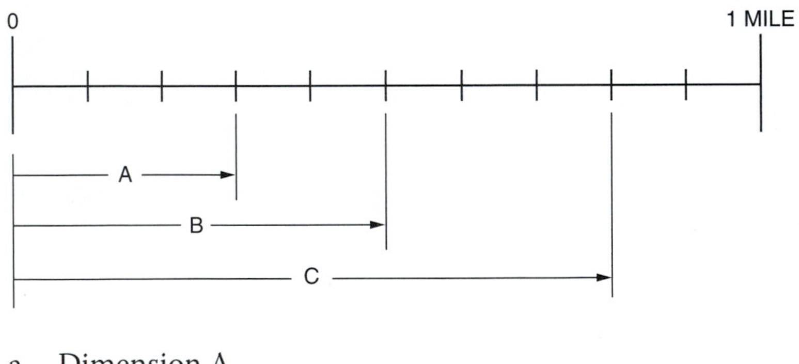

a. Dimension A　　a. ________

b. Dimension B　　b. ________

c. Dimension C　　c. ________

2. a. Round off to the nearest tenth.

   a. 10.36 = __________

   b. 4.71 = __________

   c. 0.44 = __________

   d. 10.918 = __________

   b. Round off to the nearest hundredth.

   a. 10.368 = __________

   b. 4.714 = __________

   c. 0.449 = __________

   d. 10.9189 = __________

3. Round off to the nearest whole number.

   a. 7.8 = __________

   b. 12.2 = __________

   c. 9.8 = __________

   d. 17.498 = __________

4. Express as decimal fractions.

   a. twelve hundredths a. __________

   b. seventy-eight hundredths b. __________

   c. four tenths c. __________

   d. nine tenths d. __________

   e. six-hundred twenty-seven thousandths e. __________

5. A welded tank holds 26.047 gallons. Round to the nearest hundredth of a gallon. __________

Using the Calculator

Use of the calculator in math can give you several advantages. A properly working calculator in experienced hands is 1) accurate, and 2) a timesaver.

In choosing a calculator several things are considered, such as 1) name brand, 2) cost, and 3) scientific vs basic.

Name Brand

As in all tools and equipment the craftsman needs, the name brand is there for a reason: this manufacturer usually has the experience, know-how, and resources to provide a reliable quality instrument.

Scientific vs. Non-Scientific Models

All your calculations in the field, and in this book, can be solved using the $+$, $-$, $\times$, and $\div$ functions. The possible problem with a scientific calculator is the excess quantity of buttons not normally used that can be accidently pushed or touched, complicating the process. In some models these buttons take up so much space that the actual buttons needed have to be made tinier to fit. However, scientific calculators can be used well in the classroom, some manufacturers have addressed the button size issue, and good quality name brand scientific calculators can cost as little as $10.00 to $15.00.

Many scientific calculators have the order of operations automatically programmed (see unit 25 regarding order of operations), while non-scientific calculators may not. To check your calculator for order of operation programming, do the following calculation by hand and then by calculator (answer on bottom of this page):

$$8 + 9 \times 3 - 2 \times 14 \div 7 + 3 =$$

Non-scientific calculators will accomplish all the calculations the welder would normally do in the field. It usually has only the buttons needed, and may therefore be easier to use.

Correct answer: $8 + 9 \times 3 - 2 \times 14 \div 7 + 3 \rightarrow 8 + (9 \times 3) - (2 \times 14 \div 7) + 3 = 34$

Incorrect answer: $8 + 9 \times 3 - 2 \times 14 \div 7 + 3 = 101$

Basic Calculator Use for First Time Users

EXAMPLE: 35 + 8 =

NOTE: The calculator will do exactly what you tell it to do.

1. Turn on the calculator and enter the number 35 by pressing the 3 and then the 5. The 3 first appears in the ones column, but only temporarily until you enter the 5. The 3 then moves to its proper tens column and the number 35 will display.

2. Press the + button. This lets the calculator know two things: 1) the number 35 is complete, and 2) it is going to add the next entry (or set of order of operations entries). The + sign may not appear.

3. Press the 8. The 35 may disappear, and the calculator is now waiting for your next instruction.

4. Press the = button. The calculator will do the math you have instructed it to do, 35 + 8, and display 43. It may not do that final addition calculation until you press the = button.

Regularly check the calculator viewer to make sure you have entered the correct entry. If in the previous exercise you entered 36 instead of 35, the calculator will give you 44 as an answer, as it should. It will not know you have mistakenly entered the wrong number. If you notice you entered the wrong number or told it to do the wrong function, tapping the CE or C button can erase your last entry ("C" stands for "clear" and "E" stands for "entry", or "error"). Tapping the "CE" button twice in some calculators will clear all entries. If you have made an error, it may be best to start over completely. Some calculators may have a "CA" button, which stands for "clear all".

Best Practice

Do all calculations in this book in the open spaces and not on separate pages. You will be able to keep this book as a reference for many years. Over time, if you forget a procedure, seeing your own work will go a long way in reminding you of the steps you used to solve the problem. Loose pages can and will be lost.

Notes and Helpful Hints

Calculator Hazard

Calculator overuse and over-reliance can lead to math tables memory loss and math skills loss!

Math Beginners

For your own benefit, you must be able to comfortably do all math calculations by hand. Do not use the calculator in the beginning as it may prevent you from:

1) Memorizing multiplication tables, and 2) learning important math skills. Once you are familiar with math procedures, however, practicing on the calculator will lead to accuracy in use. Check your own work as needed by doing math calculations a second or even a third time.

Some calculators may act slightly differently than described previously: practice is important.

Important Note

It's best not to take a test using an unfamiliar calculator. Some trade apprenticeship programs or job application procedures have a math entrance test, some do not. Some will allow calculators to be used, some will not. Practice on a calculator many times until you are proficient at using it correctly before bringing to an entrance test. I have occasionally seen a student fail an entrance test they could have otherwise passed because the calculator became confusing during testing. If you find yourself confused because of the calculator, discontinue using it and continue the test by hand. Remember, some math tests for apprenticeship or job entry may allow retesting, but only after a waiting period. This period could be a matter of days, weeks, or months, and some tests are given only once a year.

Calculator Exercises

Use the calculator to solve, check your work by hand, and show work below.

Symbols: (′) = feet, (″) = inches, (cm) = centimeters, (mi) = miles

Addition:

a. $15 + 9 =$ ____________

b. $381 + 16 =$ ____________

c. $61.3' + 83' =$ ____________

d. $2.375'' + 6.50'' =$ ____________

e. $0.500 + 0.125 =$ ____________

Subtraction:

f. $65 - 32 =$ ____________

g. $9 \text{ cm} - 0.246 \text{ cm} =$ ____________ cm

h. $0.062 - 0.055 =$ ____________

i. $1476 \text{ mi} - 43.9 \text{ mi} =$ ____________ mi

Multiplication:

j. $375' \times 5 =$ ____________′

k. $26.3 \times 2 =$ ____________

l. $0.306 \times 5.67 =$ ____________

m. $14.03 \times 1.11 =$ ____________

Division:

n. $35 \div 7 =$ ____________

o. $7 \div 35 =$ ____________

p. $63 \div 3.6 =$ ____________

q. $27.25 \div 0.04 =$ ____________

Solving formulas with the use of the calculator: all formulas and math problems in this book are solved by first writing down the problem, then performing step-by-step calculations.

EXAMPLE: How many feet of fencing is needed to enclose a rectangular lot $85.25' \times 30.5'$?

STEP 1 Write down the formula.

Perimeter = 2L + 2W

STEP 2 Solve by calculations.

2L = 2(85.25) On your calculator: $2 \times 85.25 = 170.5$
2W = 2(30.5) On your calculator: $2 \times 30.5 = 61$

STEP 3 Solve using formula.

On your calculator: 170.5 + 61 = 231.5′ of fencing.

Calculator Formula Exercises

r. How many square feet are in the fenced-in lot in the previous example? (See Unit 26 for formula.) r. ______________

s. How many square inches are on the surface of a circular piece of steel with a radius of 14.625″? (See Unit 31 for formula.) s. ______________

UNIT 13

Addition and Subtraction of Decimal Fractions

Basic Principles

In addition or subtraction of decimals, place the numbers so that the decimal points are lined up beneath each other, then add or subtract as you would with whole numbers.

The decimal point in the answer is also placed beneath the line of decimal points.

EXAMPLE 1: Add 10.41 + 3.6 + 14 + 31.045 + .2

STEP 1 Line up the numbers with the decimal points beneath each other.

$$\begin{array}{r} 10.41 \\ 3.6 \\ 14. \\ 31.045 \\ +\quad .2 \\ \hline . \end{array}$$

STEP 2 Use zeroes in the decimals as placeholders. Notice these zeroes do not change the value of any number as long as they are used as placeholders only and used only after the last number in the decimal.

$$\begin{array}{r} 10.410 \\ 3.600 \\ 14.000 \\ 31.045 \\ +\quad .200 \\ \hline . \end{array}$$

 Add the columns.

$$\begin{array}{r} 10.410 \\ 3.600 \\ 14.000 \\ 31.045 \\ +\ \ .200 \\ \hline . \end{array} \quad \rightarrow \quad \begin{array}{r} 10.410 \\ 3.600 \\ 14.000 \\ 31.045 \\ +\ \ .200 \\ \hline 59.255 \\ \hline \end{array}$$

ANSWER: 59.255

EXAMPLE 2: $12.5 - 6.37$

STEP 1

$$\begin{array}{r} 12.5 \\ -6.37 \\ \hline . \end{array}$$

STEP 2

$$\begin{array}{r} 12.50 \\ -6.37 \\ \hline . \end{array}$$

STEP 3

$$\begin{array}{r} {\scriptstyle 4\ 1} \\ 12.\not{5}0 \\ -6.37 \\ \hline 6.13 \end{array}$$

ANSWER: 6.13

Practical Problems

1. Solve the following:

   a. Add $3.14 + 8 + 19.6$ a. ______________

   b. Add $12.784 + .12 + 4.83$ b. ______________

   c. Add $\begin{array}{r} 65.4 \\ +\ 32.65 \\ \hline \end{array}$ c. ______________

d. Subtract 10.732 − 6 d. ______

e. Subtract 6.7 − 1.385 e. ______

f. Subtract 23.8
− .972 f. ______

2. What is dimension A? ______

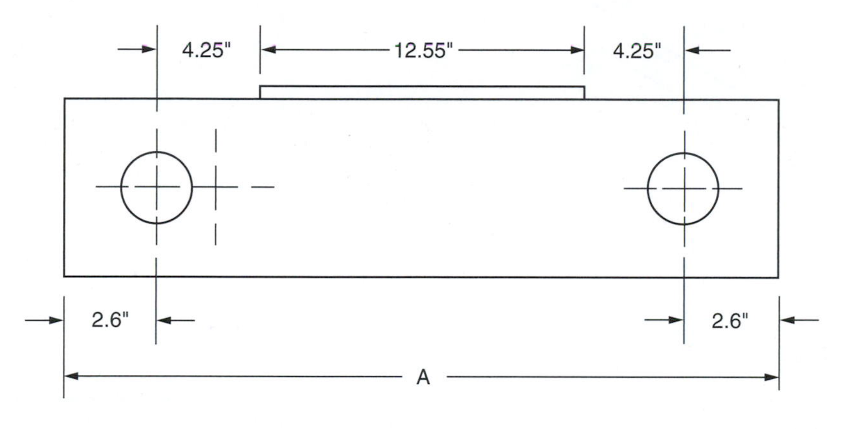

3. What is the total weight of these three pieces of steel? Round the answer to two decimal places. ______

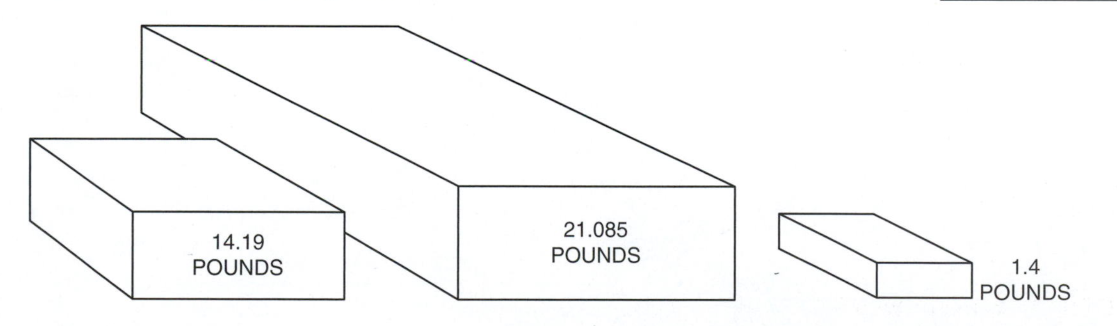

4. a. A welder saws a 3.57″ piece from this bar. Find, to the nearest hundredth inch, the length of the remaining bar. Allow a cutting loss of 0.125″. ______________

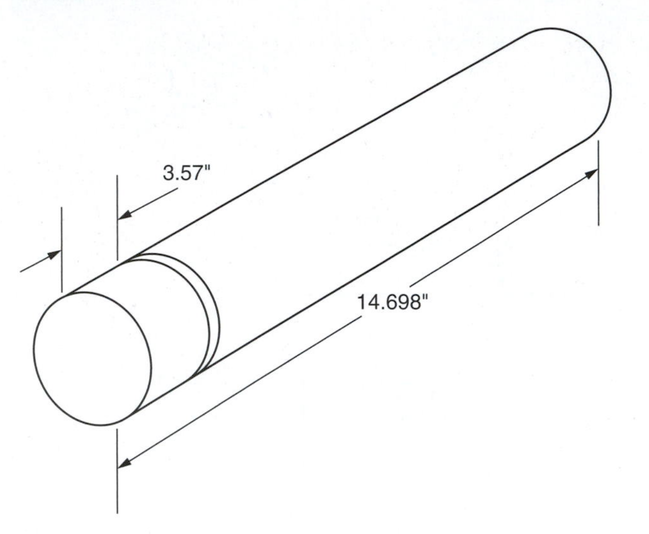

b. From the remaining bar, 2 more cuts are made: 6.45″ and 1.36″. Loss for each cut is still 0.125″. What length of bar now remains? ______________

UNIT 14

Multiplication of Decimals

Basic Principles

Decimals are multiplied in the same manner that whole numbers are multiplied. The decimal point is ignored until the final answer is calculated.

STEP 1 Set up the multiplication as you would whole numbers, ignoring the decimal point.

STEP 2 The total number of decimal places in both numbers of the problem will determine where the decimal point is placed in your answer.

EXAMPLE 1: $3.7 \times 2 =$

STEP 1 Set up and complete the multiplication.

$$\begin{array}{r} 3.7 \\ \underline{\times 2} \end{array} \quad \rightarrow \quad \begin{array}{r} \overset{1}{3}.7 \\ \underline{\times 2} \\ 74 \end{array}$$

STEP 2 Count the decimal places in both the top and bottom numbers. There is one place in the top number and none in the bottom number. The total number of decimal places is one. The decimal point is now placed one place in from the end of the answer. The arrow ◡ below shows direction and number of places the decimal point is moved.

$$\begin{array}{r} 3.7 \\ \times 2 \\ \hline 74 \end{array} \quad \rightarrow \quad \rightarrow \quad \begin{array}{r} {}^{1} \\ 3.7 \\ \times 2 \\ \hline 7.4. \end{array}$$

ANSWER: $3.7 \times 2 = 7.4$

EXAMPLE 2: $3.25 + 4.5$

STEP 1

$$\begin{array}{r} 3.25 \\ \times \;\; 4.5 \\ \hline \end{array} \quad \rightarrow \quad \begin{array}{r} {}^{1\;2} \\ {}^{\not{1}\;2} \\ 3.25 \\ \times \;\; 4.5 \\ \hline 1625 \\ 1300 \\ \hline 14625 \end{array}$$

STEP 2 Count the decimal places in both numbers of the problem.

3.25 has two places. (.25)

4.5 has one place. (.5)

There are a total of three decimal places.

Place the decimal point three places in from the end.

$$\begin{array}{r} 3.25 \\ \times \;\; 4.5 \\ \hline 1625 \\ 1300 \\ \hline 14625 \end{array} \quad \rightarrow \quad 14.625.$$

ANSWER: $3.25 \times 4.5 = 14.625$

Exercises

Multiply the following:

1. 36×1.5 ________
2. $.03 \times 6.25$ ________

Practical Problems

1. Nineteen pipe flanges are flame-cut from 16.25″ wide plate. How much plate is required for the 19 flanges? Assume that there is no waste due to cutting. ________

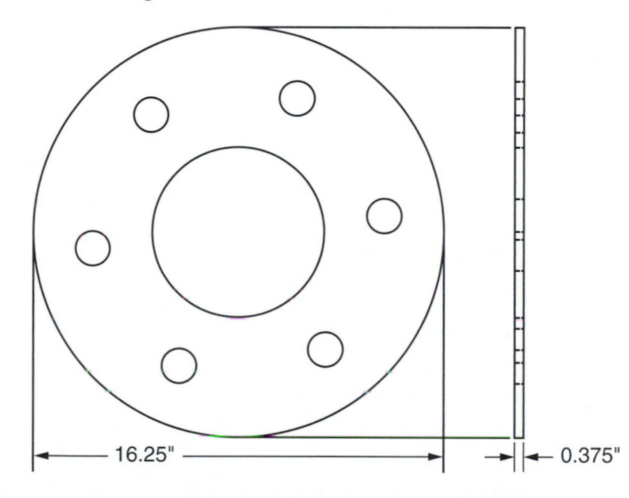

2. These 19 flanges are stacked as shown. Each flange is 0.375″ high. What is the total height of the pile? ________

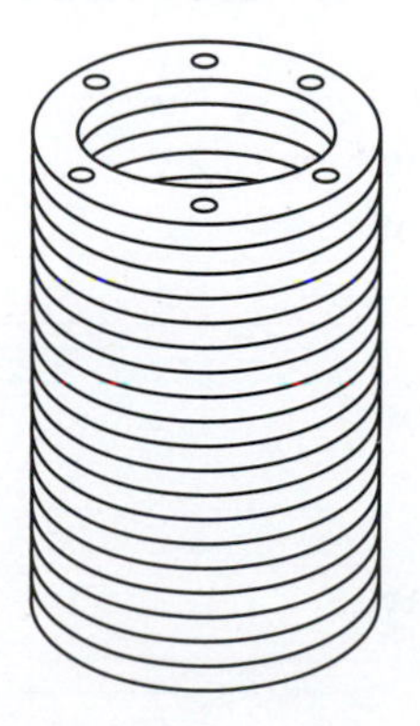

3. A welder uses 3.18 cubic feet of acetylene gas to cut one flange. How much acetylene gas is used to cut the 19 flanges? ______

4. A welder uses 7.25 cubic feet of oxygen gas to cut one flange. How much oxygen gas is used to cut 19 flanges? ______

NOTE: Use this diagram for Problems 5–7.

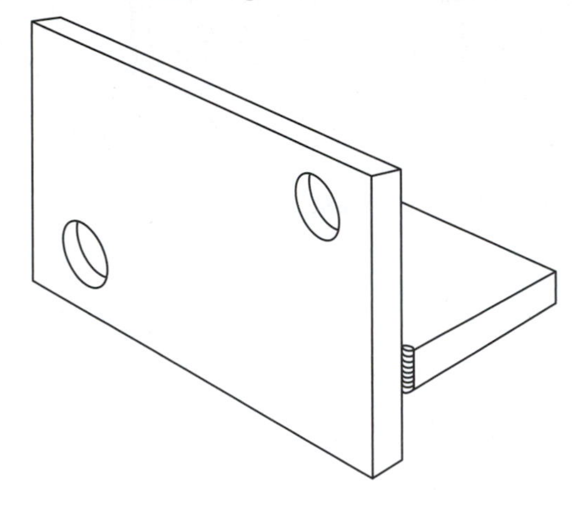

5. Each of these welded brackets weighs 2.8 pounds. A welder makes 13 of the brackets. What is the total weight of the 13 brackets? ______

6. The steel plate used to make the brackets cost $1.53 per pound. Each bracket weighs 2.8 pounds. What is the total cost of the order of 13 brackets? Round the answer to the nearest whole cent. ______

7. The welder cuts two holes in each bracket. Each bolt-hole cut wastes 0.1875 pound of material. Find, in pounds, the amount of waste for the order of 13 brackets. ______

8. a. A welder cuts 14 squares from a piece of plate. Each side is 4.125″. What is the total length of 4.125″-wide stock needed? Round the answer to two decimal places. Disregard waste caused by the width of the cuts. ______

   b. From a 96″ length of 4.125″ stock, how many inches are used for the 14 squares if the kerf width is 0.125″? ______

9. A MIG unit has a melt-off of 1.6 pounds/hour of wire. How many pounds will be melted in 16.25 hours? ______

UNIT 15

Division of Decimals

Basic Principles

Decimals are divided in the same manner that whole numbers are divided.

RULE: The divisor, the number doing the dividing, must be a whole number. A decimal can be made into a whole number by moving the decimal point to the end of the number.

EXAMPLE 1: 22.4 divided by 3 (22.4 ÷ 3)

$3\overline{)22.4}$

STEP 1 The divisor (3) is a whole number; therefore, no changes are needed.

STEP 2 Bring the decimal point straight up from its current position onto the division box. The decimal point is now in its correct place.

$$3\overline{)22.4} \rightarrow \overset{.}{3\overline{)22.4}}$$

STEP 3 For rounding to hundredths, add 0s to the dividend, as needed for three places.

$$\overset{.}{3\overline{)22.4}} \rightarrow \overset{.}{3\overline{)22.400}}$$

STEP 4 Divide as in normal division.

```
    7.466     →   7.466 rounded to hundredths = 7.47
3)22.400
  21
   14
   12
    20
    18
     2
```

ANSWER: 22.4 ÷ 3 = 7.47

Calculator Reminder

The steps used to solve math problems by calculator or by hand are similar. The difference is that either you do the math yourself, or the calculator does the work.

EXAMPLE 1: 22.4 ÷ 3 = ____________

Calculator Steps

Enter the problem into the calculator by pushing buttons.

STEP 1 Push 22.4.

STEP 2 Push "÷".

STEP 3 Push "3".

STEP 4 Push "="; answer shows on display screen.

STEP 5 Round off to hundredths.

EXAMPLE 2: 11.73 divided by 1.2 (11.73 ÷ 1.2)

$1.2\overline{)11.73}$

STEP 1 Move the divisor's decimal point to the end of the number. This makes it a whole number.

$1.2\overline{)11.73}$ $12\overline{)11.73}$

STEP 1a The decimal point in the dividend must also be moved to the right the same number of places as the decimal point in the divisor was moved. (1 place in this example.)

$12\overline{)11.73}$ $12\overline{)117.3}$

STEP 2 Bring the decimal point straight up from its new position onto the division box. The decimal point is now in its correct place.

$$\begin{array}{r} . \\ 12\overline{)117.3} \end{array}$$

STEP 3 For rounding to hundredths, add 0s to the dividend, as needed.

$$\begin{array}{r} . \\ 12\overline{)117.300} \end{array}$$

STEP 4 Divide as in normal division.

$$\begin{array}{r} 9.775 \\ 12\overline{)117.300} \\ \underline{-108} \\ 93 \\ \underline{-84} \\ 90 \\ \underline{-84} \\ 60 \\ \underline{-60} \\ 0 \end{array}$$

9.775 = 9.78 rounded to hundredths

ANSWER: 11.73 ÷ 1.2 = 9.78

Practical Problems

1. a. A welder saw cuts this length of steel angle into 7 equal pieces. What is the length of each piece? Disregard waste caused by the width of the cuts. Round the answer to 2 decimal places. __________

   b. What is each piece measurement if the angle is cut into 9 equal lengths? Kerf width is 0.125. __________

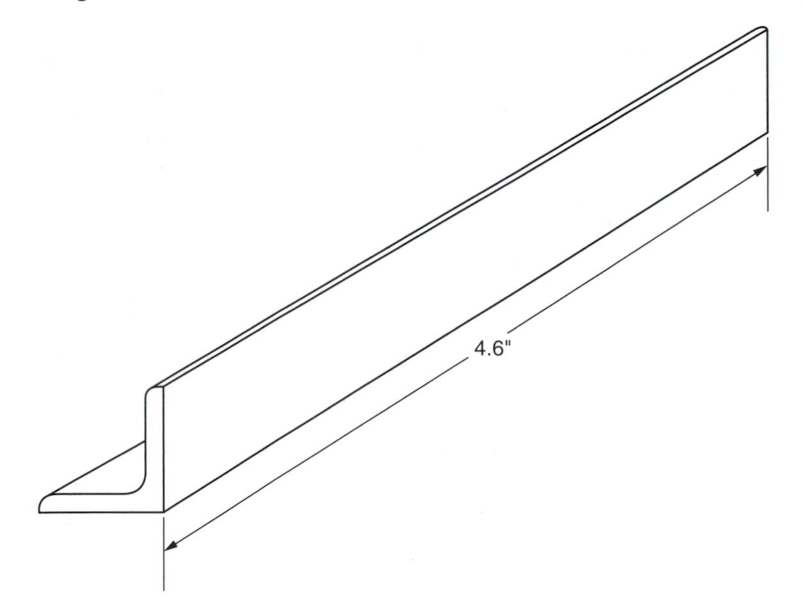

2. A welder cuts this plate into pieces that are 9.25″ wide. How many whole pieces are cut? Disregard waste caused by the width of the cuts (a). a. __________

   How many pieces are cut if kerf width is .625 (b)? b. __________

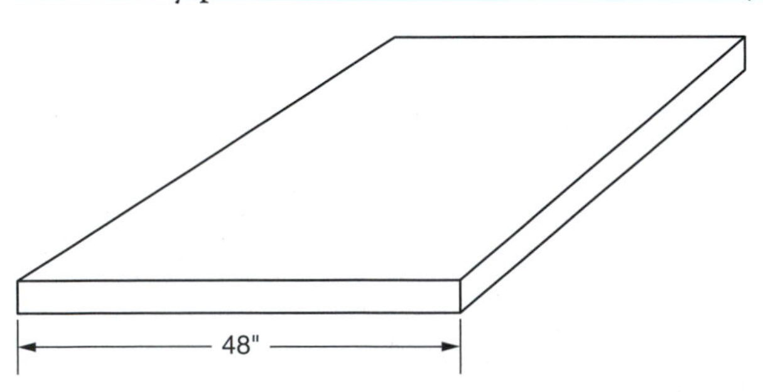

3. A welder shears key stock into pieces 2.75″ long. How many whole pieces are sheared from a length of key stock 74.15″ long? ______

4. The welder saws this square stock into 4 equal pieces. What is the length of each piece? Allow 1⁄16″ waste for each cut. Round your answer to the nearest hundredth. ______

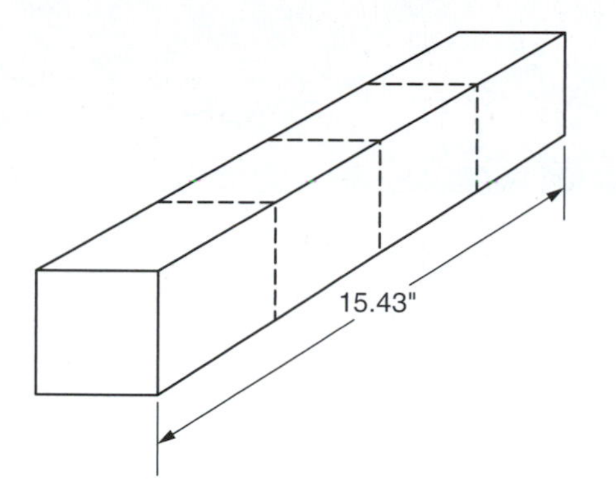

5. a. The welder flame-cuts this plate into 7 equal pieces, each 12″ long. Find, to the nearest hundredth inch, the width of each piece. Allow 3⁄16″ waste for each cut. ______

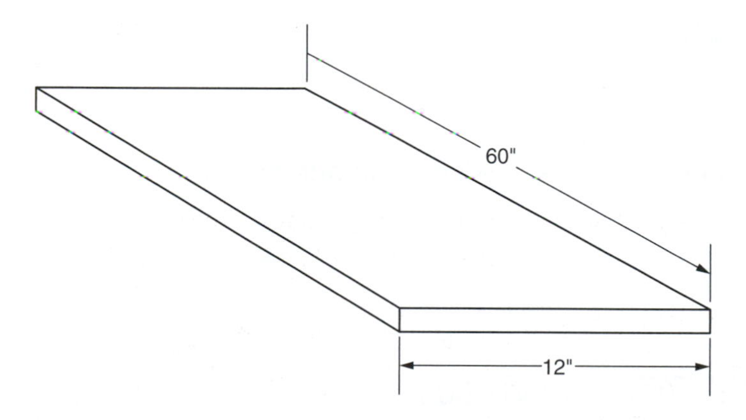

b. How many full strips 1.75″ × 60″ can be cut, allowing a 1/8″ cut width? ______

c. What width of 60″ scrap is left over? ______

6. A MIG unit is using a 33-pound spool of wire. How many hours will the spool last if the melt-off is 2.35 pounds/hour? How many hours will a 500-pound drum last? ______ ______

UNIT 16

Decimal Fractions and Common Fraction Equivalents

Basic Principles

Change Fractions to Decimals

RULE: Divide the numerator (top) by the denominator (bottom).

EXAMPLE 1: Change ¼ into a decimal.

STEP 1 Set up the problem.

$4\overline{)1}$

STEP 2 Place the decimal point in the dividend (the number 1), and add 0s.

$4\overline{)1.000}$

STEP 3 Divide.

$$4\overline{)1.000} \rightarrow \overset{.}{4\overline{)1.000}} \rightarrow \begin{array}{r} .25 \\ 4\overline{)1.000} \\ \underline{8} \\ 20 \\ \underline{20} \\ 0 \end{array}$$

ANSWER: $\frac{1}{4} = .25$

EXAMPLE 2: Change 16⅜″ into a decimal.

RULE: Only the fractional part of a number (⅜″) is changed.

STEP 1 Divide the numerator by the denominator.

$8\overline{)3}$

STEP 2 $8\overline{)3.000}$

STEP 3

$$8\overline{)3.000} \rightarrow 8\overline{)\dot{3}.000} \rightarrow \begin{array}{r} .375 \\ 8\overline{)3.000} \\ \underline{24} \\ 60 \\ \underline{56} \\ 40 \\ \underline{40} \\ 0 \end{array}$$

= .375. Keep answer at three places if it ends with 5.

ANSWER: $16\frac{3}{8}'' = 16.375$

Change Decimals into Fractions

METHOD 1: The number of decimal places determines the number of 0s used in the denominator with the number 1. The decimal number itself becomes the numerator.

EXAMPLE 1: Change .3 into a fraction.

STEP .3 has one decimal place, so the denominator has one 0.

ANSWER: $.3 = \frac{3}{10}$

RULE: The decimal point is not transferred to the numerator.

EXAMPLE 2: Change 4.75 into a fraction.

RULE: Only the fractional part of the number (.75) is changed.

STEP .75 has two decimal places, so the denominator has two 0s.

$$.75 = \frac{75}{100} \quad \frac{75}{100} \text{ reduces to } \frac{3}{4}$$

ANSWER: $4.75 = 4\frac{3}{4}$

METHOD 2: Speak the decimal out loud correctly.

EXAMPLE: Change .37 into a fraction.

STEP .37 said out loud is "thirty-seven hundredths".

ANSWER: $.37 = \frac{37}{100}$

Practical Problems

1. Express the fractional inches as a decimal number.

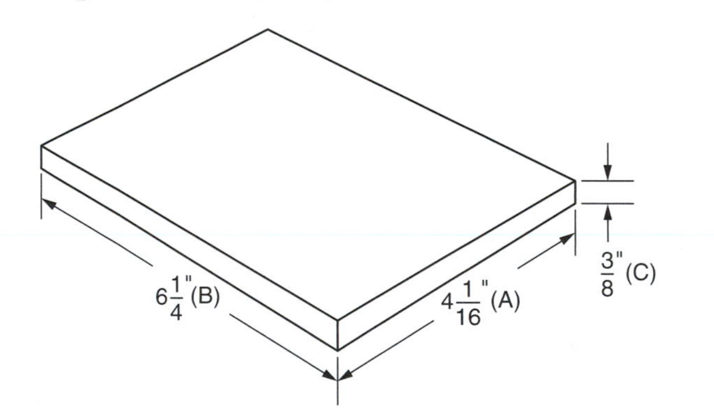

a. Dimension A a. ______

b. Dimension B b. ______

c. Dimension C c. ______

2. Express each dimension in feet and inches. Express the fractional inches as a decimal number.

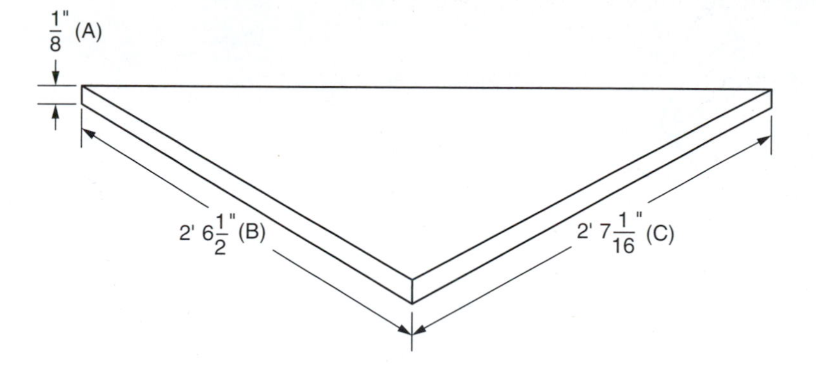

a. Dimension A a. ______________

b. Dimension B b. ______________

c. Dimension C c. ______________

3. Express each decimal dimension as a fractional number.

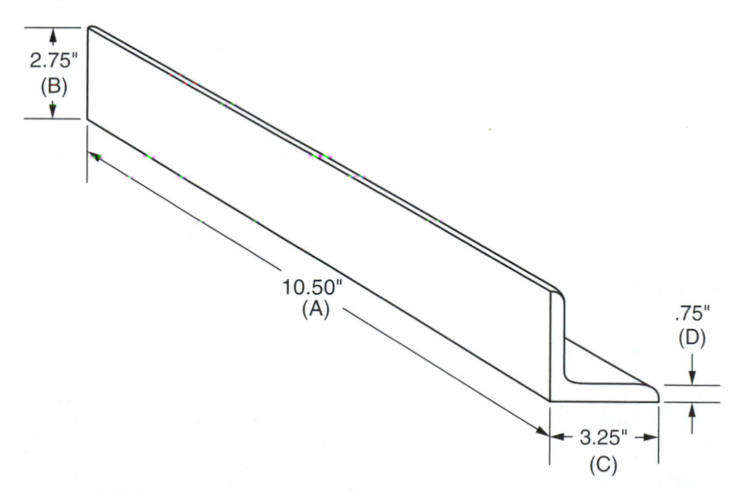

a. Dimension A a. ______________

b. Dimension B b. ______________

c. Dimension C c. ______________

d. Dimension D d. ______________

4. Express each dimension as a fractional number.

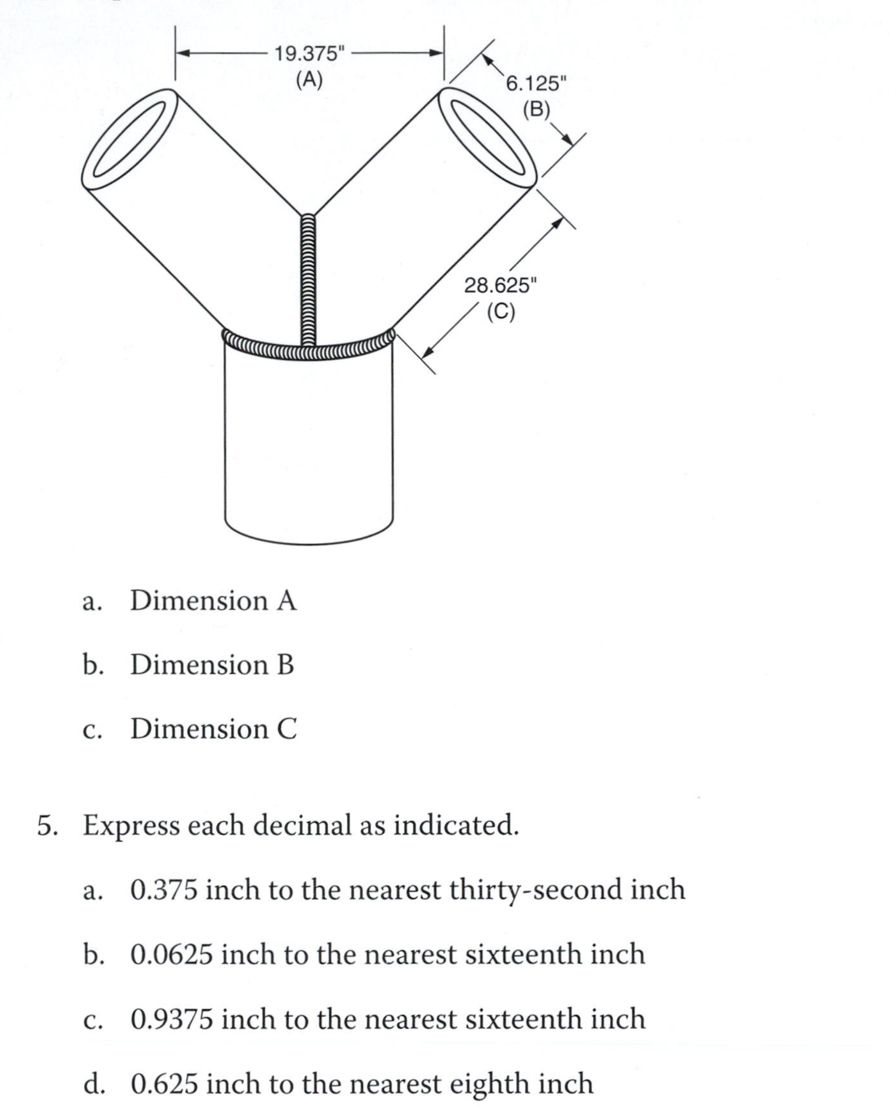

a. Dimension A — a. ______

b. Dimension B — b. ______

c. Dimension C — c. ______

5. Express each decimal as indicated.

a. 0.375 inch to the nearest thirty-second inch — a. ______

b. 0.0625 inch to the nearest sixteenth inch — b. ______

c. 0.9375 inch to the nearest sixteenth inch — c. ______

d. 0.625 inch to the nearest eighth inch — d. ______

e. 0.750 inch to the nearest fourth inch — e. ______

See the following Example.

RULE: Multiply the decimal by the denominator asked for. That answer becomes the numerator of the fraction.

EXAMPLE: .125 to thirty-seconds: $.125 \times 32 = 4$

$$.125 = \frac{4}{32}$$

6. A welder cuts these four pieces of metal. Express each dimension as a fractional number.

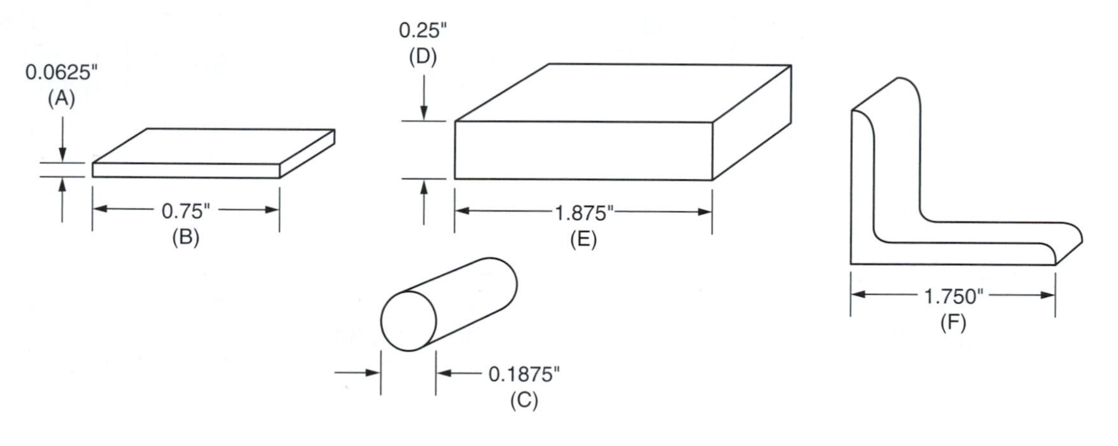

a. Dimension A a. ______

b. Dimension B b. ______

c. Dimension C c. ______

d. Dimension D d. ______

e. Dimension E e. ______

f. Dimension F f. ______

NOTE: Use this table for Problems 8a–f

DECIMAL EQUIVALENT TABLE

Fraction	Decimal	Fraction	Decimal	Fraction	Decimal	Fraction	Decimal
1/64	0.015625	17/64	0.265625	33/64	0.515625	49/64	0.765625
1/32	0.03125	9/32	0.28125	17/32	0.53125	25/32	0.78125
3/64	0.46875	19/64	0.296875	35/64	0.546875	51/64	0.796875
1/16	0.0625	5/16	0.3125	9/16	0.5625	13/16	0.8125
5/64	0.078125	21/64	0.328125	37/64	0.578125	53/64	0.828125
3/32	0.09375	11/32	0.34375	19/32	0.59375	27/32	0.84375
7/64	0.109375	23/64	0.359375	39/64	0.609375	55/64	0.859375
1/8	0.125	3/8	0.375	5/8	0.625	7/8	0.875
9/64	0.140625	25/64	0.390625	41/64	0.640625	57/64	0.890625
5/32	0.15625	13/32	0.40625	21/32	0.65625	29/32	0.90625
11/64	0.171875	27/64	0.421875	43/64	0.671875	59/64	0.921875
3/16	0.1875	7/16	0.4375	11/16	0.6875	15/16	0.9375
13/64	0.203125	29/64	0.453125	45/64	0.703125	61/64	0.953125
7/32	0.21875	15/32	0.46875	23/32	0.71875	31/32	0.96875
15/64	0.234375	31/64	0.484375	47/64	0.734375	63/64	0.984375
1/4	0.250	1/2	0.500	3/4	0.750	1	1.000

7. A piece of steel channel and a piece of I beam are needed. Express each dimension as a decimal number.

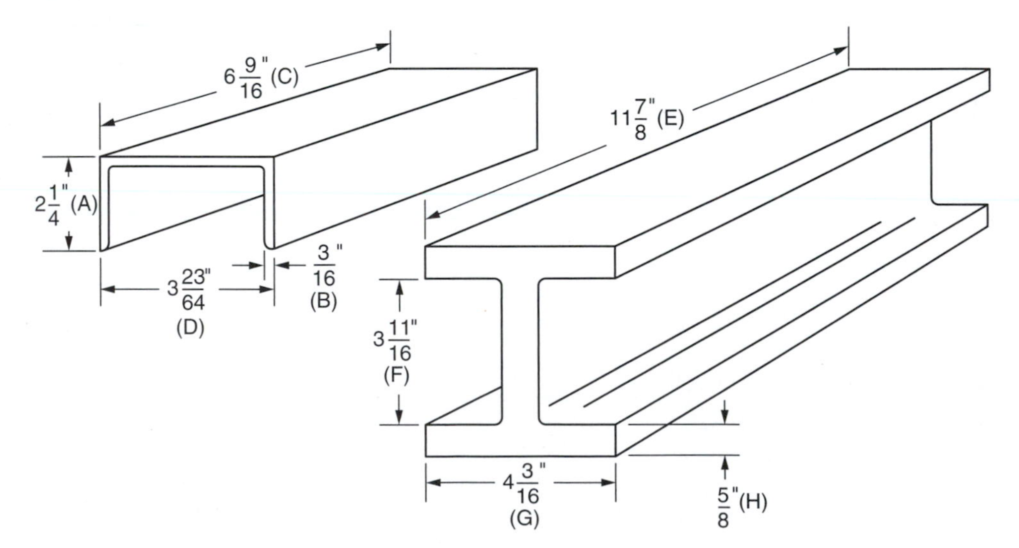

a. Dimension A — a. ________

b. Dimension B — b. ________

c. Dimension C — c. ________

d. Dimension D — d. ________

e. Dimension E — e. ________

f. Dimension F — f. ________

g. Dimension G — g. ________

h. Dimension H — h. ________

8. Express each decimal as a fraction. Use the decimal equivalent table to answer 8 a–f.

a. 0.171875 — a. ________

b. 0.3125 — b. ________

c. 0.875 — c. ________

d. 0.5625 — d. ________

e. 0.9375 — e. ________

f. 0.515625 — f. ________

UNIT 17

Tolerances

Basic Prinicples

Tolerance is the allowable amount greater or lesser than a given measurement.

EXAMPLE: Lengths of 0.75″-diameter shafts are used in the manufacture of certain tools. The manufacturer needs the shafts to be 10.55″ long, yet he and the design engineer have determined that if the shafts are a little longer or shorter than 10.55″, the tools work just as well and are safe. The engineer calculates that the shafts cannot be less than 10.50″ in length, nor greater than 10.60″ in length.

The measurement on the blueprint reads 10.55″ plus or minus .05″ ($10.55''^{\pm .05}$).

To determine the greatest length, A:

$$\begin{array}{r} 10.55 \\ +\ .05 \\ \hline 10.60'' \end{array} \qquad A = 10.60''$$

To determine the shortest length, B:

$$\begin{array}{r} 10.55 \\ -\ .05 \\ \hline 10.60'' \end{array} \qquad B = 10.50''$$

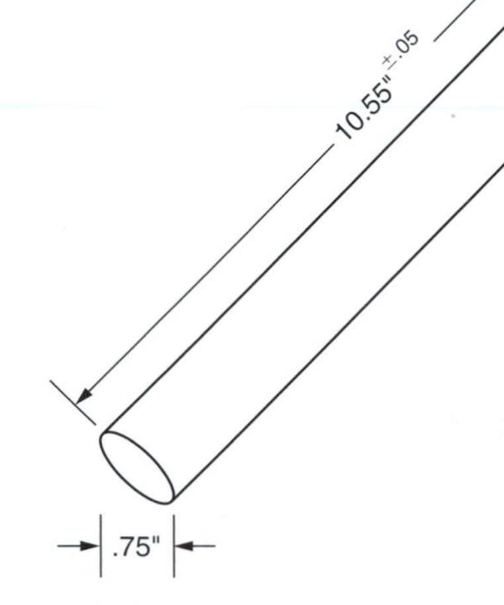

The tolerance for the diameter of this shaft will also be shown on the blueprint. Tolerance of the diameter may be more critical than the tolerance of the length.

Reminder: ∅ symbol = diameter

Practical Problems

1. Find the allowed minimum and maximum diameter of the shaft if the ∅ is given as such:

   ∅ = $.75^{\pm .005}$

   A. Maximum ∅ ____________

   B. Minimum ∅ ____________

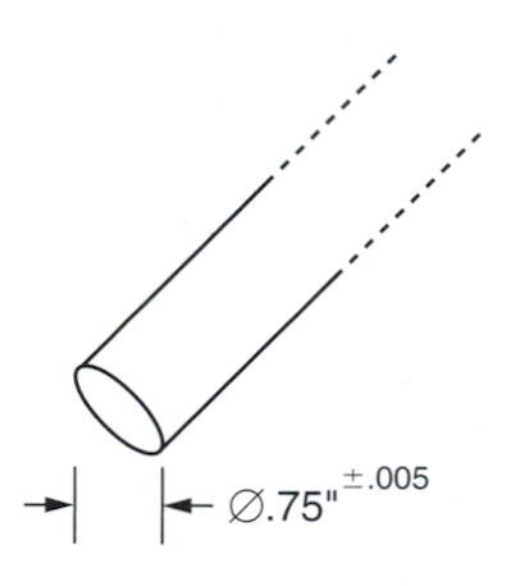

2. Using the given tolerances, find A, the largest allowable measurement, and B, the smallest allowable measurement.

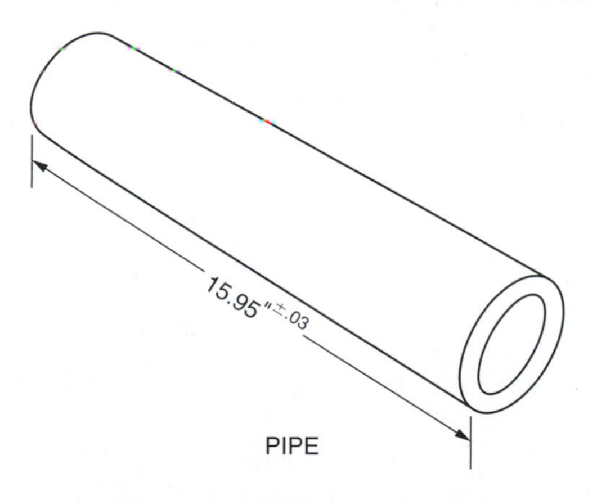

A = ____________

B = ____________

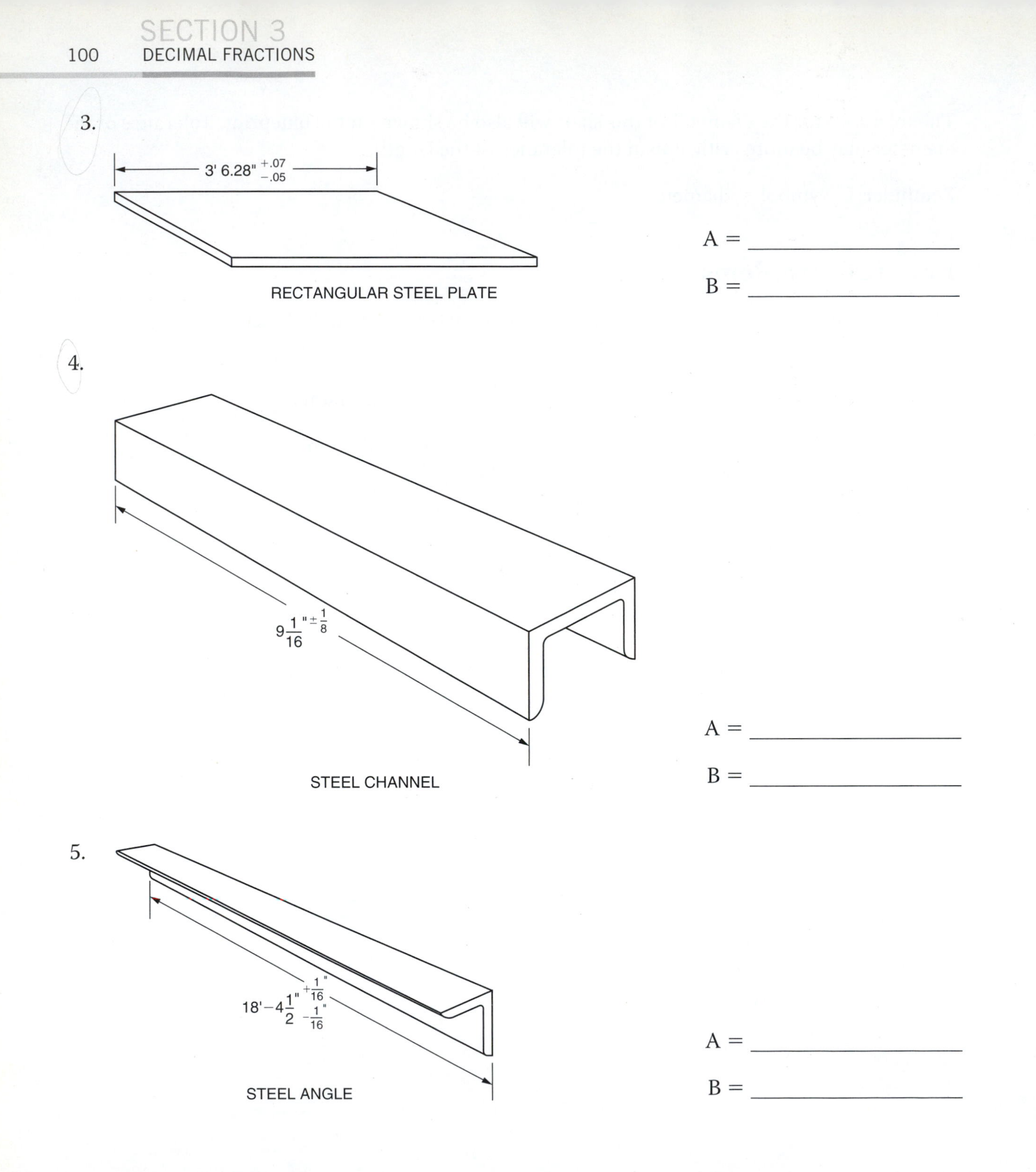
3.
3' 6.28" +.07 −.05
RECTANGULAR STEEL PLATE
A = ____________
B = ____________
4.
9 1/16" ± 1/8
STEEL CHANNEL
A = ____________
B = ____________
5.
18'−4 1/2" +1/16" −1/16"
STEEL ANGLE
A = ____________
B = ____________

6.

7.

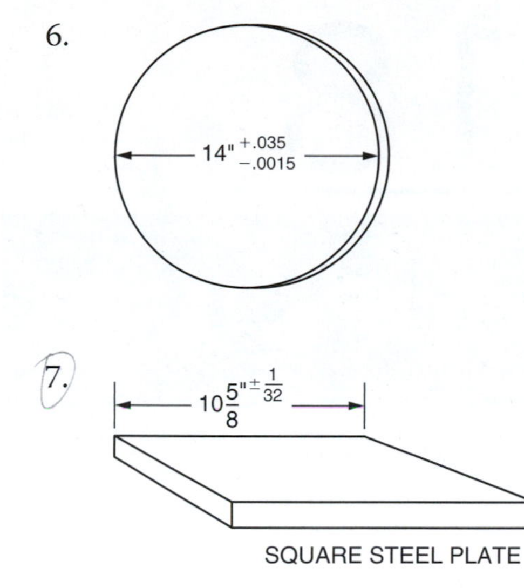

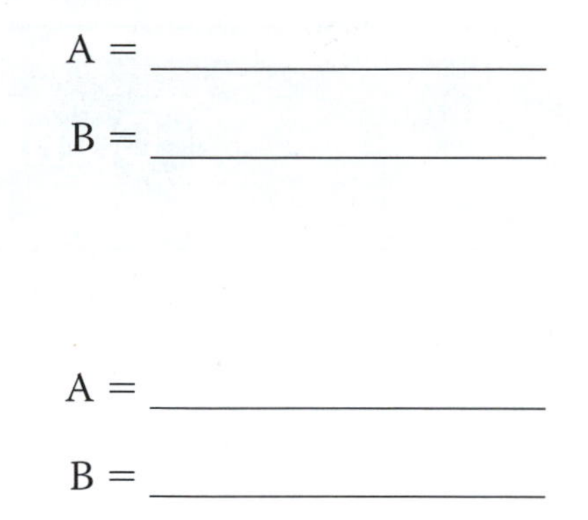

UNIT 18

Combined Operations with Decimal Fractions

Basic Principles

Solve the problems in this unit using addition, subtraction, multiplication, and division of decimals.

EXAMPLE: If A = 3.25″, C = 3.25″, and the total width of A,B, and C = 8.5625″, find the diameter of B.

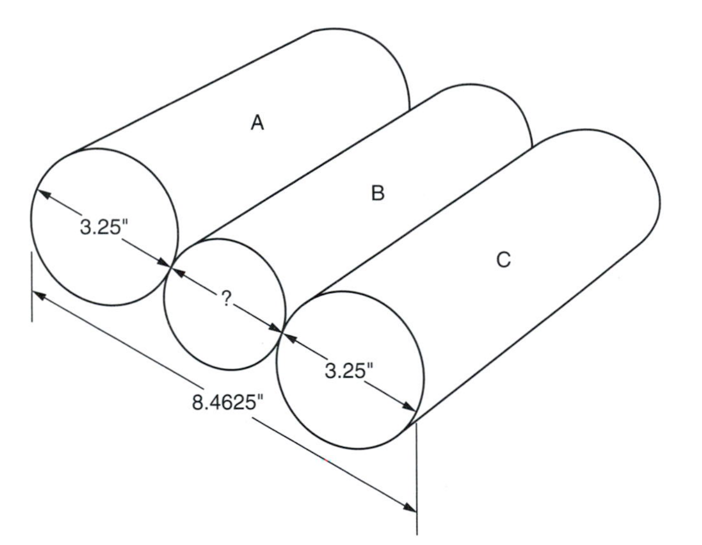

$$
\begin{array}{rl}
3.25 & = \varnothing A \\
+3.25 & = \underline{\varnothing C} \\
\hline
6.50 &
\end{array}
$$

Then:

$$
\begin{array}{rl}
8.4625 & \text{Total width} \\
-6.50 & \\
\hline
1.9625'' & = \varnothing B
\end{array}
$$

To proof the work, add all 3 diameters together.

$$
\begin{array}{r}
3.25 \\
3.25 \\
1.9625 \\
\hline
8.4625''
\end{array}
$$

Practical Problems

1. Find the length of slot 2. ______________

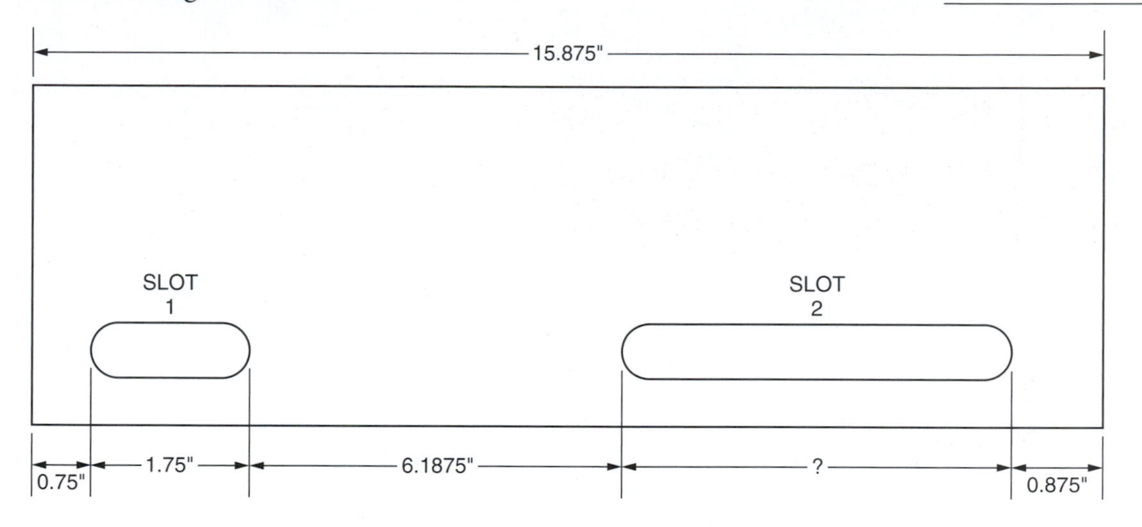

2. Find the height of a stack containing 13 of these steel shims. ______________

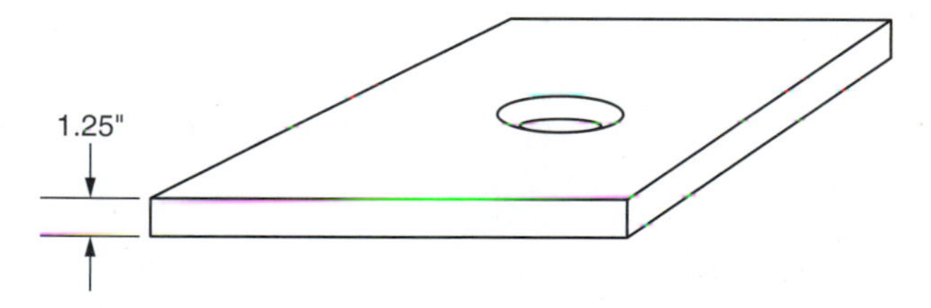

3. Cross-bar members are cut from flat stock. What length of 5″ flat stock is used to make 31 of these members? Disregard waste caused by the width of the cuts. ______________

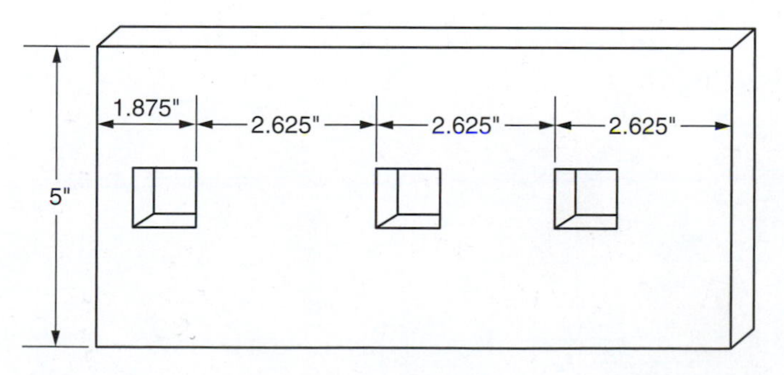

4. The round stock shown here is cut into 3.5″ pieces. How many pieces can be made? Allow 0.125″ waste for each cut. ____________

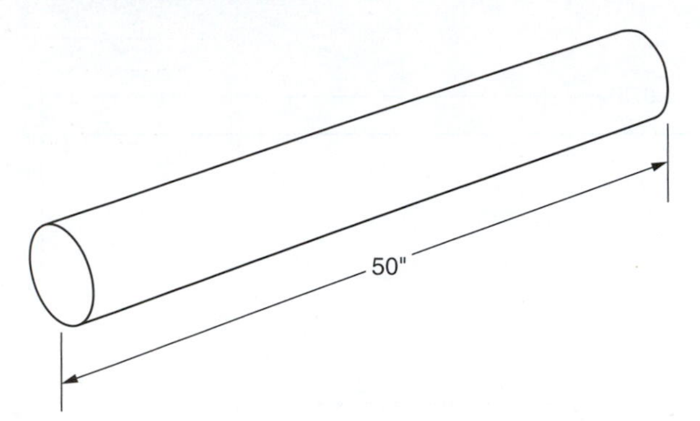

5. A welded truck-bed side support is shown. How many complete supports can be cut from a length of 2″ × 127.875″-long plate stock? Disregard waste caused by the width of the cuts (a). Allow .135 waste for each cut (b).

a. ____________

b. ____________

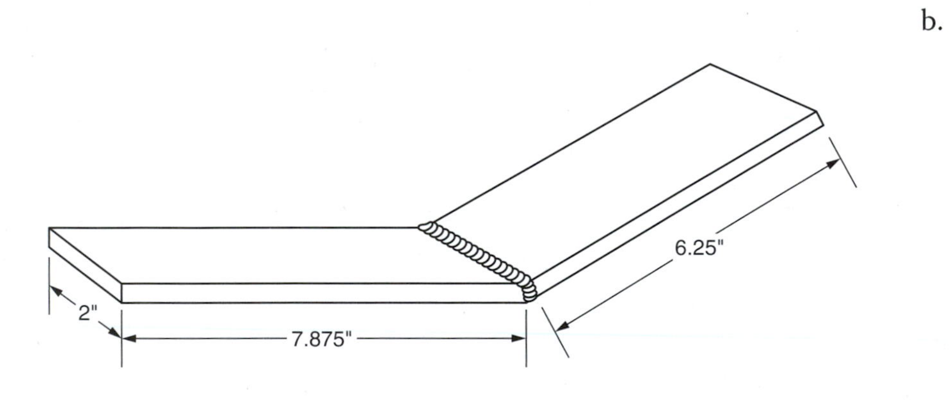

6. Round off the length of this steel channel to three decimal places; round off the length to hundredths; round off the length to tenths. ____________

____

____

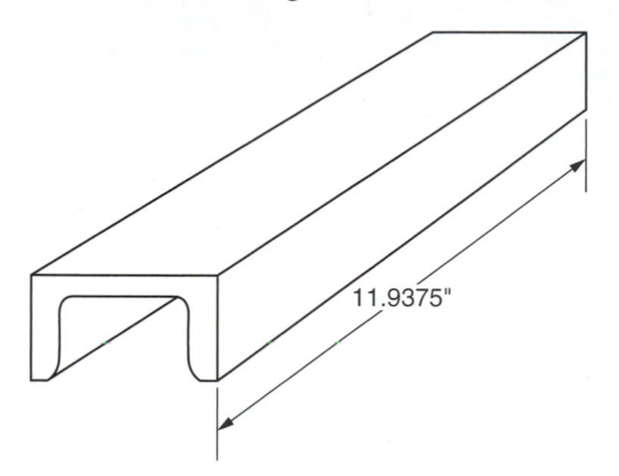

7. Find the minimum and maximum allowable diameters of the flange, and the minimum and maximum distance allowable from the center of the flange to the bolt-hole circle.

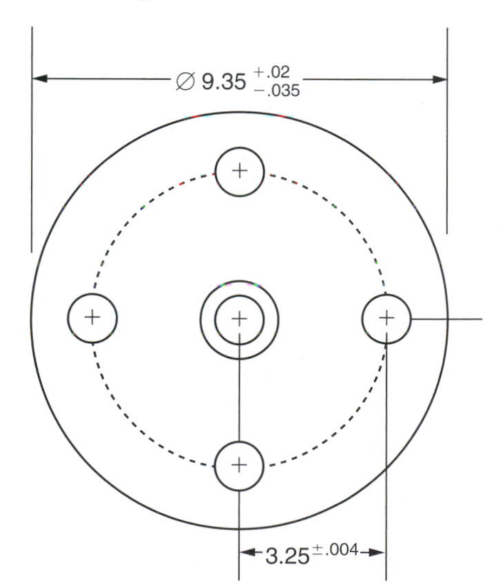

flange: a. ____________

b. ____________

bolt-hole circle: a. ____________

b. ____________

8. Convert each fraction to a decimal. Add the answers together for total, e.

a. $\frac{3}{4}$ a. ____________

b. $\frac{17}{32}$ b. ____________

c. $\frac{5}{16}$ c. ____________

d. $\frac{7}{8}$ d. ____________

e. ____________

9. What is the thickness of the wall of a pipe that has an inside diameter of 15.72 inches and an outside diameter of 16.50 inches? ____________

10. A welded bracket has the dimensions shown. Find dimension A. ____________

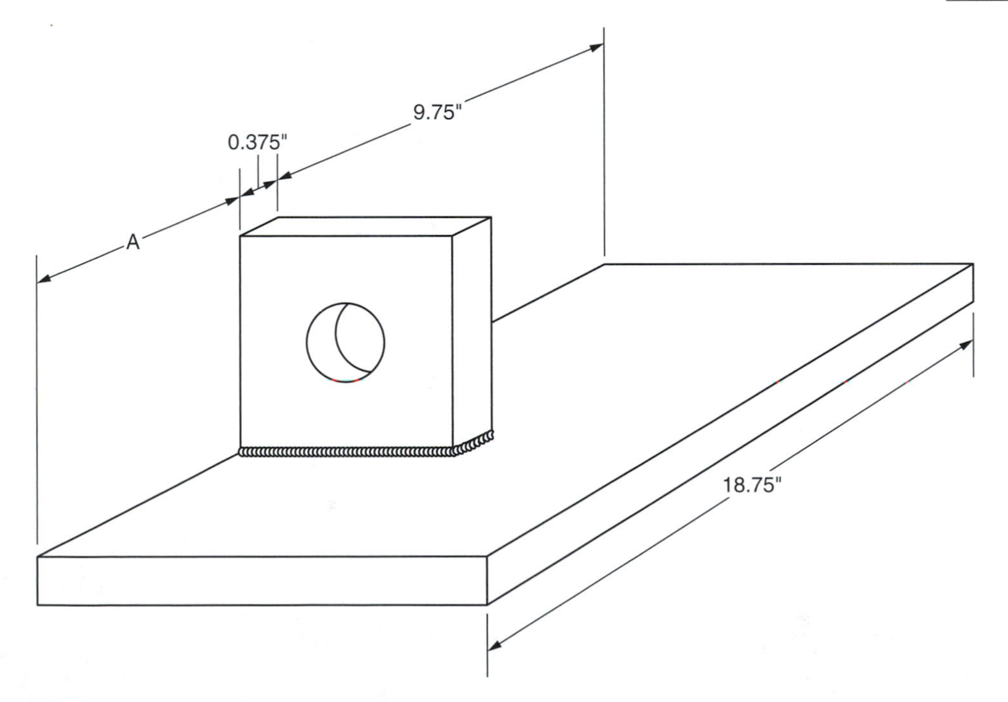

UNIT 19

Equivalent Measurements

Basic Principles

Study this table of equivalent units.

ENGLISH LENGTH MEASURE

1 foot (ft)	=	12 inches (in)
1 yard (yd)	=	3 feet (ft)
1 mile (m)	=	1,760 yards (yd)
1 mile (m)	=	5,280 feet (ft)

EXAMPLE: A length of steel angle is 8′ long. Express this measurement in inches.

QUESTION: How many inches are there in 8′?

SOLUTION: Each foot has 12″:

$$\text{multiply} \quad \begin{array}{r} 12 \\ \times\ 8 \\ \hline \end{array} \rightarrow \begin{array}{r} 12 \\ \times\ 8 \\ \hline 96 \end{array} \rightarrow 96''$$

ANSWER: 8′ = 96″

Practical Problems

1. Express this 23′ length of the steel channel in inches only. ________

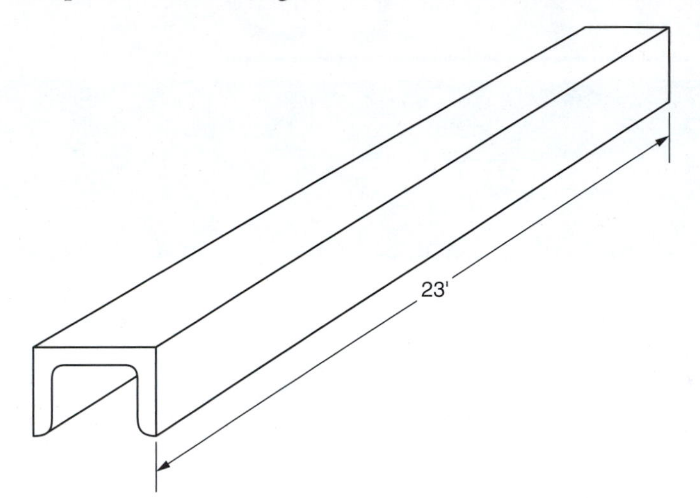

2. Express the distance between hole centers in feet only. ________

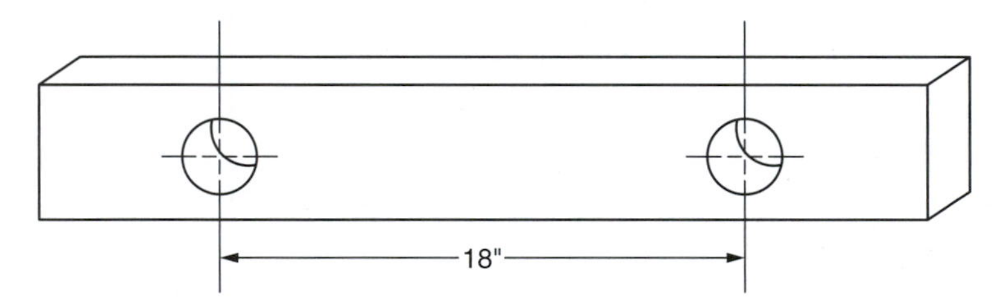

3. Express this length of the round stock shown in feet. ________

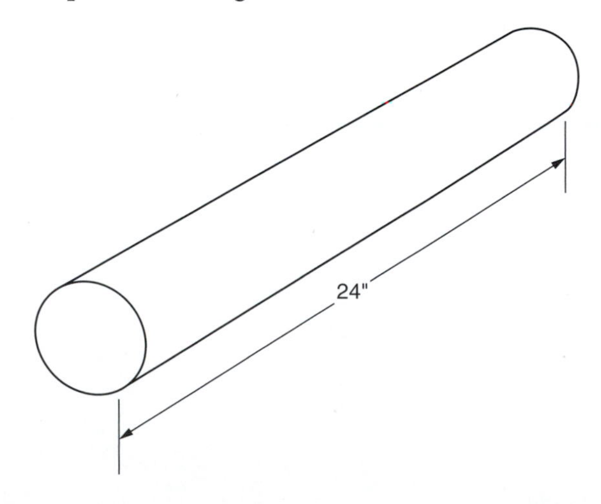

4. Express the length of this pipe in inches. ____________

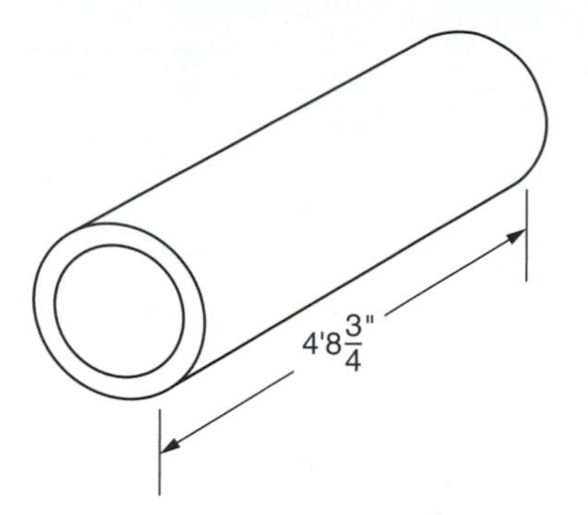

5. The fillet weld shown has 3′, plus 18″ of weld on the other side of the joint. Express the total amount of weld in feet. ____________

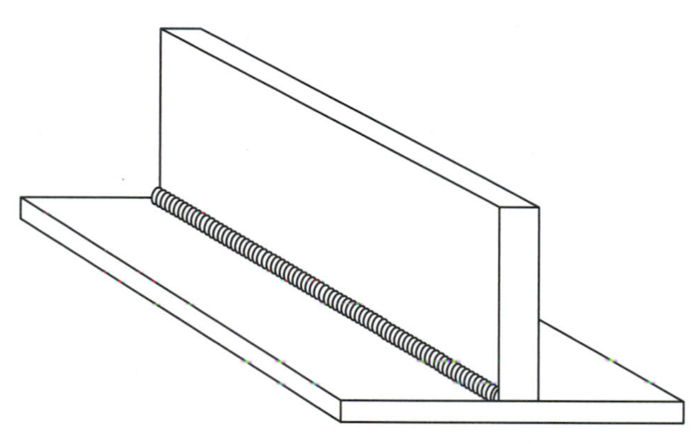

6. This tee-bracket is 18.625″ long. Express the measurement in feet, inches, and a fractional part of an inch. ____________

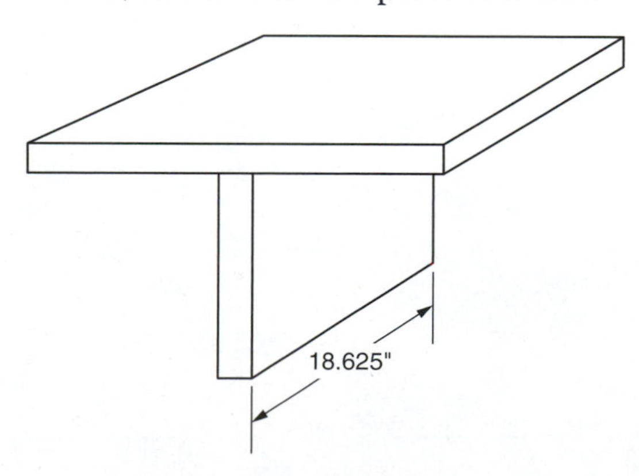

7. A piece of I beam is 3′ 2.75″ long. Express this measurement in inches only. ____________

8. A circular steel plate is flame-cut and the center is removed.

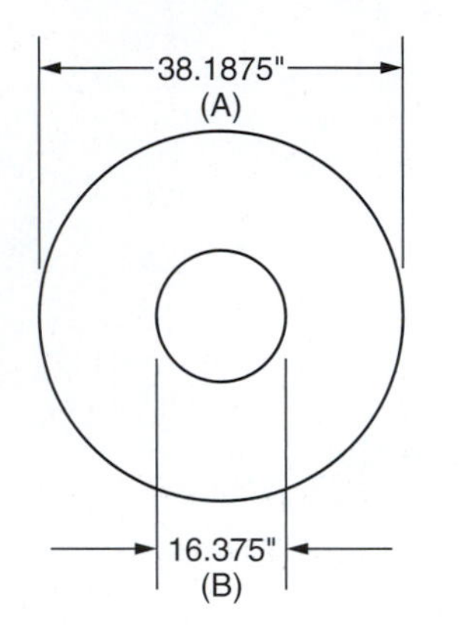

a. Express diameter A in feet, inches, and a fractional part of an inch. a. ____________

b. Express diameter B in feet, inches, and a fractional part of an inch. b. ____________

SECTION 4

AVERAGES, PERCENTAGES, AND MULTIPLIERS

UNIT 20

Averages

Basic Principles

Averaging figures can give the welder helpful information about a variety of subjects.

EXAMPLE: Find the average weekly pay for Jim, a welder at Smith Steel, during the month of March.

Week 1, Jim made $784.00
Week 2, $631.00
Week 3, $815.00
Week 4, $736.00

RULE: To find the average of two or more figures, first they are added together; the sum is then divided by the number of figures.

STEP 1 Add the figures together.

$$\begin{array}{r} 784 \\ 631 \\ 815 \\ +736 \\ \hline \end{array} \quad \rightarrow \quad \begin{array}{r} \scriptstyle 1\,1 \\ 784 \\ 631 \\ 815 \\ +736 \\ \hline 2966 \end{array}$$

 Divide the sum by the number of figures.

```
                741             741.50
4)2966  →  4)2966   →  4)2966.00
               28            28
                16            16
                16            16
                 6             6
                 4             4
                 2            20
                              20
                               0
```

ANSWER: Jim averaged $741.50 per week in March.

Practical Problems

1. Find the average number of miles driven per day using the following figures:

   Monday: 73 miles ______________

   Tuesday: 28 miles

   Wednesday: 136 miles

   Thursday: 61 miles

   Friday: 48 miles

   Saturday: 59 miles

2. Five pieces of 1″ square bar stock are cut as shown. What is the average length, in inches, of the pieces? __________

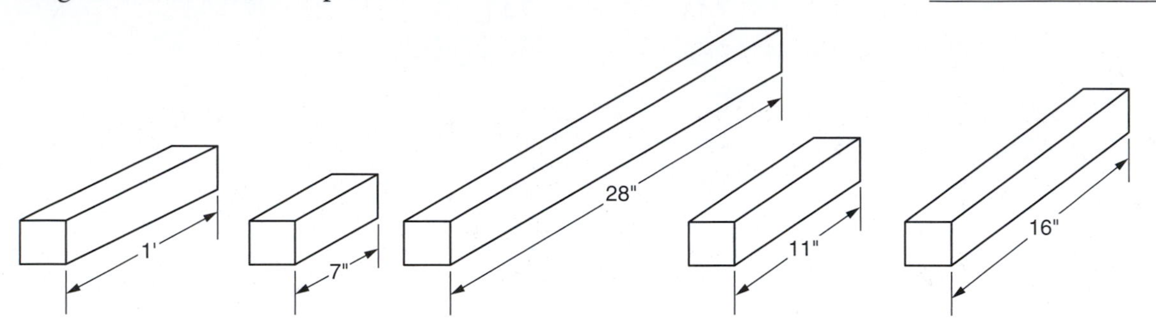

3. Six welding jobs are completed using 33 pounds, 19 pounds, 48 pounds, 14 pounds, 31 pounds, and 95 pounds of electrodes. What is the average poundage of electrodes used for each job? __________

4. On 5 jobs, a welder charges 6.5 hours, 3 hours, 11 hours, 2.75 hours, and 9.25 hours. Find the average hours billed per job. __________

5. Four pieces of steel plate are measured for thickness. These measurements are found:

$$1\frac{1}{4}'', 1\frac{3}{16}'', 1\frac{1}{4}'', 1\frac{1}{2}''$$

What is the average thickness of the plate? Round the answer to the nearest thousandth. __________

6. Four pieces of ½″ plate weigh 10.85 pounds, 26 pounds, 9¾ pounds, and 29½ pounds. Find the average weight of the plates to the nearest thousandth pound. __________

7. Six plates are stacked and weighed. The total weight is 210 pounds. What is the average weight of each piece? __________

8. A welded steel tank holds 325 gallons. Another tank holds twice as much. What is the average amount held by the tanks? __________

UNIT 21

Percents and Percentages (%)

Basic Principles

Percents are used to express a part or a portion of a whole. They are based on the principle that 100% represents a whole, 50% represents one-half, 25% represents one-quarter, etc.

EXAMPLE 1: A customer places an order with your company for 40 welded brackets.

A. When 20 brackets are completed and pass inspection, 50% of the order is finished.

B. When all 40 are completed, 100% of the order is finished.

PROCEDURE: To calculate percentages (%) as in the above:

a. Formulate a fraction from the information given.

b. Change the fraction into a decimal.

c. Change the decimal into a %.

A. To calculate the % when 20 brackets are completed:

STEP 1 Formulate a fraction from the information given.

$\frac{20}{40}$ "20 of 40" brackets are completed. $\frac{20}{40}$ reduces to $\frac{1}{2}$.

STEP 2 Change the fraction to a decimal.

$$\frac{1}{2} \rightarrow 2\overline{)1} \rightarrow \begin{array}{r} .50 \\ 2\overline{)1.00} \\ \underline{10} \\ 00 \end{array}$$

$$\frac{1}{2} = .50$$

STEP 3 Change the decimal into a %.

PROCEDURE:

a. move the decimal point two places to the right, and

b. put the % sign at the end of the number.

$.50 \rightarrow .50. \rightarrow 50.\% = 50\%$

The decimal point is not shown if it is at the end of the number.

ANSWER: When 20 of 40 brackets are completed, 50% of the order is finished.

B. To calculate the % when all 40 of the brackets are completed:

STEP 1 Formulate a fraction from the information given.

$\frac{40}{40}$ "40 of 40" brackets are completed.

$\frac{40}{40}$ reduces to 1.

STEP 2 Change the fraction to a decimal.

In this case, the fraction reduced to the whole number 1. The decimal now needs to be shown.

$1 \rightarrow 1.$

STEP 3 Move the decimal point two places to the right, and put the % sign at the end.

1 → 1. . 100. → 100.% → 100%

ANSWER: When 40 of 40 brackets are completed, 100% of the order is finished.

EXAMPLE 2: Jim has driven 17 of the 25 miles he travels to his worksite. What % of the drive has he completed?

STEP 1 $\frac{17}{25}$ "17 of 25" miles

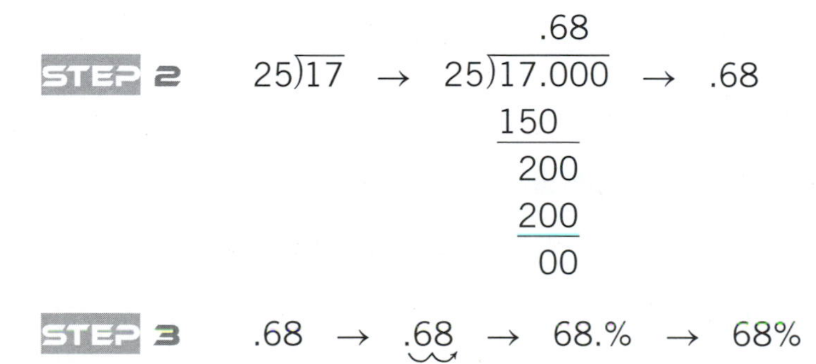

STEP 2 25)17 → 25)17.000 → .68

.68
150
200
200
00

STEP 3 .68 → .68 → 68.% → 68%

ANSWER: Jim has completed 68% of his commute.

PROCEDURE: ***To change a % to a decimal:***

a. move the decimal point two places to the left, and

b. remove the % sign.

EXAMPLE 1: Change 12.4% to a decimal.

a. 12.4% → .12.4% → .124%

b. .124% → .124

ANSWER: 12.4% = .124

EXAMPLE 2: Change 4% to a decimal.

a. 4% → 4.% → .4.% → .04%

b. .04% → .04

ANSWER: 4% = .04

PROCEDURE: ***The procedure for computing a percentage of any given number is to:***

a. Change the % into a decimal.

b. Multiply the given number by that decimal.

EXAMPLE: The usual retail price of an oxy-acetylene cutting outfit is $485.00. However, there is a 7.5% discount the day you buy it.

a. How much will you save with the 7.5% discount? (What is 7.5% of $485.00?)

STEP 1 Change 7.5% into a decimal.

7.5% → .7.5% → .075

STEP 2 Multiply $485.00 × .075.

```
 485          485
×.075   →    ×.075
              2425
             3395
             36375  →  36375.  →  36.375
```

$36.375 rounds off to $36.38.

a. Savings: $36.38

b. What is the price of the cutting outfit after the discount?

$$\begin{array}{r} 485.00 \\ -\ \ 36.38 \\ \hline \end{array} \rightarrow \begin{array}{r} \scriptstyle 7\,14\;\,9\,1 \\ 4\not{8}\not{5}.\not{0}0 \\ -\ \ 36.38 \\ \hline \$448.62 \end{array}$$

b. Price after discount: $448.62

c. If the tax rate is 12%, how much tax has to be paid?

STEP 1

12% → 12.% → 12.% → .12

$$\begin{array}{r} 448.62 \\ \times \quad .12 \\ \hline \end{array} \rightarrow \begin{array}{r} \scriptstyle 1\,1\quad\;\; \\ 448.62 \\ \times \quad .12 \\ \hline 89724 \\ 44862 \\ \hline 538344. \end{array}$$

$53.8344 = $53.83

c. Tax: $53.83

d. What is your final cost?

$$\begin{array}{r} \$448.62 \\ +\quad 53.83 \\ \hline \end{array} \rightarrow \begin{array}{r} \scriptstyle 1\,1\,1\quad\;\; \\ \$448.62 \\ +\quad 53.83 \\ \hline 502.45 \end{array}$$

d. Final cost: $502.45

Practical Problems

1. Express each percent as a decimal.

   a. 16% a. ______

   b. 5% b. ______

   c. .8% c. ______

   d. 60½% d. ______

   e. 23.25% e. ______

   f. 125% f. ______

   g. 220% g. ______

2. A welder works 40 hours and earns $22.50 per hour. The deductions are:

   income tax, 18%
   Social Security, 7%
   union dues, 5%
   hospitalization insurance, 2.5%

   Find each amount to the nearest whole cent.

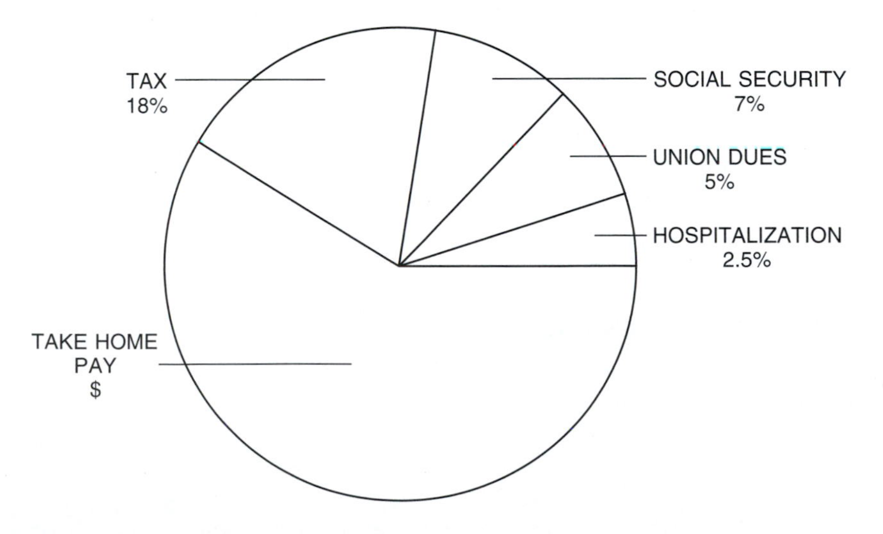

A. What is the welder's gross pay before deductions? A. ____________

Calculate:

a. hospitalization insurance a. ____________

b. income tax b. ____________

c. Social Security c. ____________

d. union dues d. ____________

e. net, or take home, pay e. ____________

3. The area of a piece of steel is 1446.45 square inches. How many square inches are contained in 25% of the steel? ____________

4. A welder completes 87% of 220 welds. How many completed welds are made? ____________

5. A total of 80 coupons (weld test plates) are submitted for certification and all are inspected. Use the information given to calculate a.–c.

TOTAL PLATES SUBMITTED	80
PASSED VISUAL	77
PASSED X-RAY AND VISUAL	68

a. What % passed visual inspection? a. ____________

b. What % passed x-ray examination? b. ____________

c. What % are inspected? c. ____________

6. In a mill, 10,206 steel plates are sheared. By inspection, 20% of the plates are rejected. Of the amount rejected, 8% are scrapped.

   a. How many plates are rejected? a. ________________

   b. How many of the rejected plates are scrapped? b. ________________

7. Develop the fraction, decimal, or % as needed.

	Fraction	Decimal	%
Example:	$\frac{1}{4}$	.25	25%
a.	$\frac{3}{8}$	____	____
b.	____	____	80%
c.	____	2.125	____
d.	$\frac{7}{32}$	____	____
e.	____	.75	____
f.	$\frac{16}{16}$	____	____
g.	____	____	100%

Multipliers and Discounts

A multiplier is used to determine a discount offered to customers by the supplier. It is used in an effort to attract business and compete with other suppliers. It is similar to percentages in that a .95 multiplier determines that a customer will pay only 95% of the stated cost of material.

EXAMPLE: Huff Steel offers a .90 multiplier on orders of product at $1,000.00, while ABC steel offers a .94 multiplier. Compare the cost of a $1,000.00 order from each company.

Huff: $1,000 × .9 = 9,000 = $900

ABC: $1,000 × .94 = 94,000 = $940.00

Huff is cheaper by $40.00.

However, if Huff charges $150.00 to ship the material, and ABC charges $100.00, what is the final cost comparison?

Huff: $900.00 + 150.00 = $1,050.00

ABC: $940.00 + 100.00 = $1,040.00

ABC is now cheaper by $10.00.

NOTE: Other factors can affect the decision-making process when selecting a supplier. These factors include:

- Order completion date reliability
- Product quality reliability
- Adherence to order specifications
- Value of having a business relationship with an individual supplier versus multiple suppliers

Practical Problems

8. Huff Steel offers a .875 multiplier on orders of $1,500.00 or more and charges $225.00 for shipping. TD Mfg. offers a .90 multiplier with $200.00 shipping charges. Which company offers lower landed material costs, and what is that cost? __________

9. CTD Steel has a "2-ten net-30" offer for its' customers. This offer allows a 2% discount if the customer pays the bill within 10 days; otherwise, the full price is due within 30 days. If a customer takes advantage of the 2-ten offer, how much can they save on a $12,378.00 order? __________

10. Steel USA uses a multiplier of .90 for its' customers on orders greater than $4,500.00. They also include an additional 2-10 net-30 offer on the bill. Find the actual materials cost on an initial order of $7,868.27 if paid in 8 days. ________________

SECTION 5

METRIC SYSTEM MEASUREMENTS

UNIT 22

The Metric System of Measurements

Basic Principles

The metric system is a set of measurements developed in the 1790s, primarily by the French, in an effort to "obtain uniformity in measures, weights, and coins . . ."

Thomas Jefferson was an early proponent of the use of the system, and the United States was the first country to develop its coinage based on metrics: our dollar is divided into 100 cents.

Prior to metrics, English measure consisted of a multitude of measurements, some of with which we are familiar; others have meanings that are antiquated.

EXAMPLES:

inch	ell
foot	furlong
yard	pole or perch
mile	fathom
hand	league

Although the United States still uses a mixture of English and metrics, most other countries in the world use only the metric system.

The **meter** is the standard unit of measurement of length in the metric system. It is several inches longer than the English "yard."

Metric:

1 METER
100 CENTIMETERS
(3.28 FEET)
10 20 30 40 50 cm 60 70 80 90 100 cm

English:

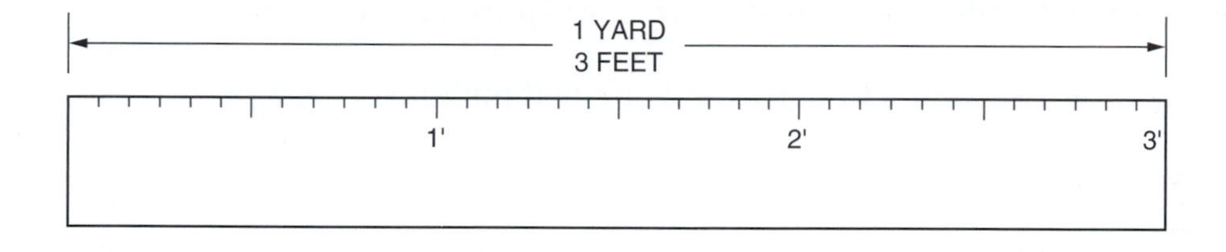

The meter is divided into 100 small units, called centimeters.

*centi*meter: "$\frac{1}{100}$ of a meter"

Root word: *centi* $\left(\frac{1}{100}\right)$

Each centimeter is divided further into 10 smaller units, called millimeters.

*milli*meter: "$\frac{1}{1,000}$ of a meter"

Root word: *milli* $\left(\frac{1}{1,000}\right)$

Commonly used metric conversions that are smaller than a meter.

1 meter (m)	=	100 centimeters (cm)
1 meter (m)	=	1,000 millimeters (mm)
1 centimeter (cm)	=	10 millimeters (mm)

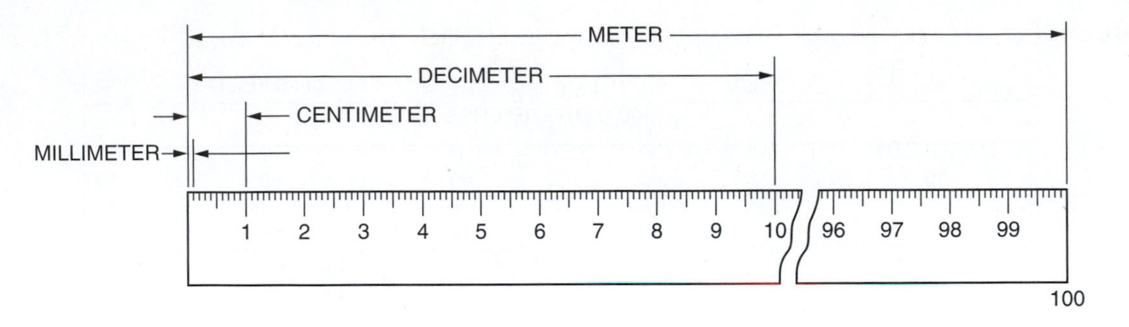

If a measurement is smaller than a millimeter, it is expressed as a decimal.

EXAMPLE: A measurement of two and one-half millimeters is written:

2.5 mm

Commonly used metric conversion that is larger than a meter:

1 kilometer (km)	=	1,000 meters (m)

*kilo*meter: "1,000 meters"

Root word: *kilo* (1,000)

Less commonly used metric conversions:

decimeter (dm) = 10 centimeters

dekameter (dam) = 10 meters

hectometer (hm) = 100 meters

Other standard units of measure in the metric system:

gram: measure of mass

liter: measure of volume

PROCEDURE:

To express large metric units in smaller units, multiply the given number by 10, 100, or 1,000 as needed.

EXAMPLE: This bar is 7.3 centimeters long. What is the length in millimeters? See the metric length measure table.

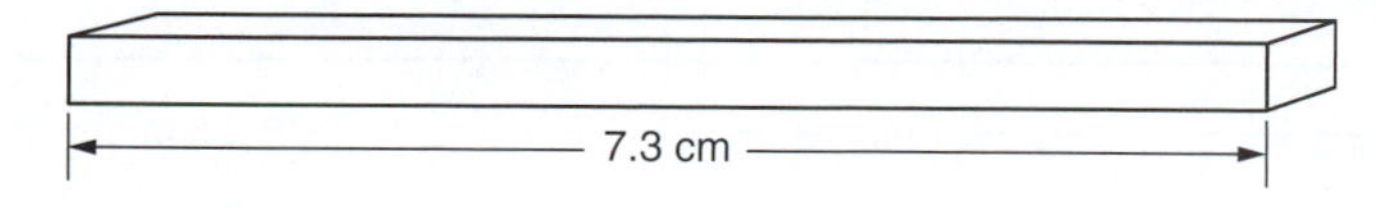

SOLUTION: Multiply by 10. Since the measurement is in centimeters, each 1 cm contains 10 millimeters.

$$\begin{array}{r} 7.3 \\ \times \quad 10 \end{array} \rightarrow \begin{array}{r} 7.3 \\ \times \quad 10 \\ \hline 73.0 \end{array} \rightarrow 73 \text{ millimeters}$$

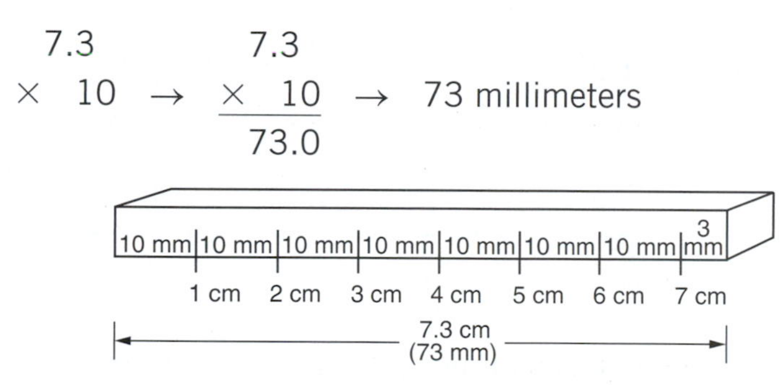

PROCEDURE:

To express small metric units in larger units, divide the given number by 10, 100, or 1,000 as needed.

EXAMPLE:

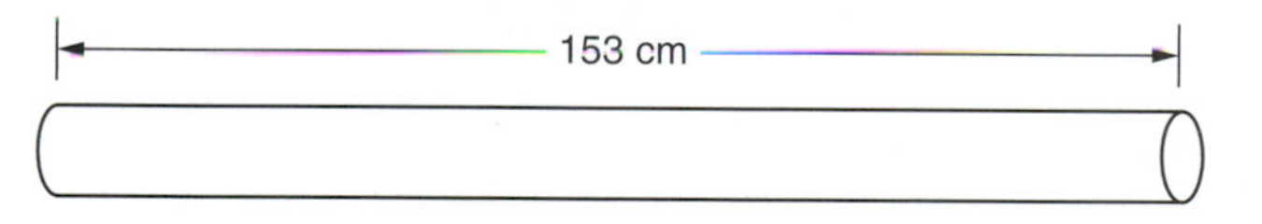

The illustrated rod is 153 cm long. What is the length in meters?

SOLUTION: Divide by 100. Since the measurement is in centimeters, each group of 100 cm equals 1 meter.

$$100\overline{)153} \rightarrow \begin{array}{r} 1.53 \\ 100\overline{)153.00} \\ \underline{100} \\ 53\,0 \\ \underline{50\,0} \\ 3\,00 \\ \underline{3\,00} \end{array} \rightarrow 1.53 \text{ m}$$

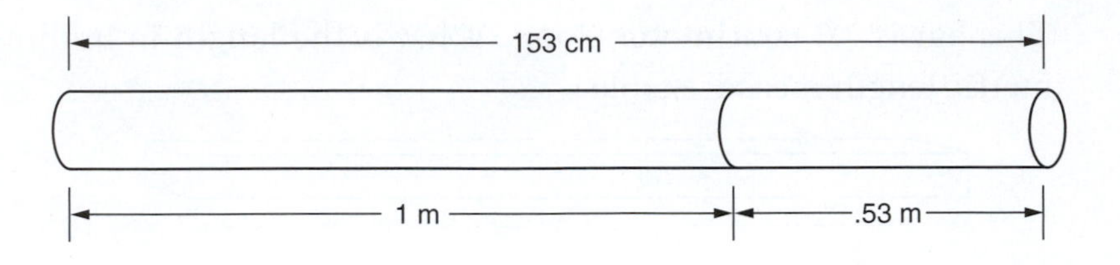

Practical Problems

NOTE: Use this diagram for Problem 1.

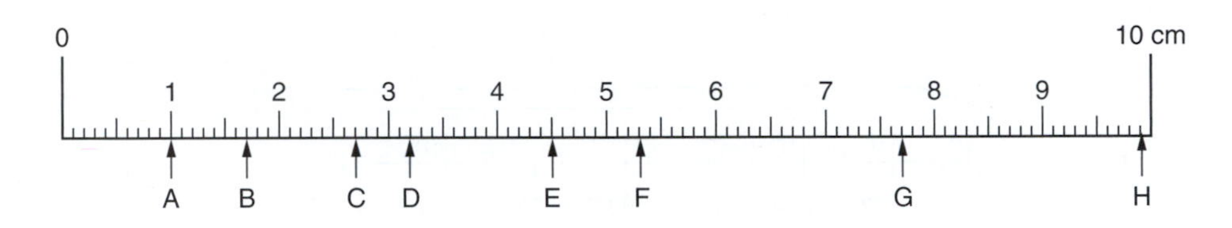

1. Read the distances, in millimeters and then in centimeters, from the start of the rule to the letters A–H on the rule. Record the answers in the proper blanks.

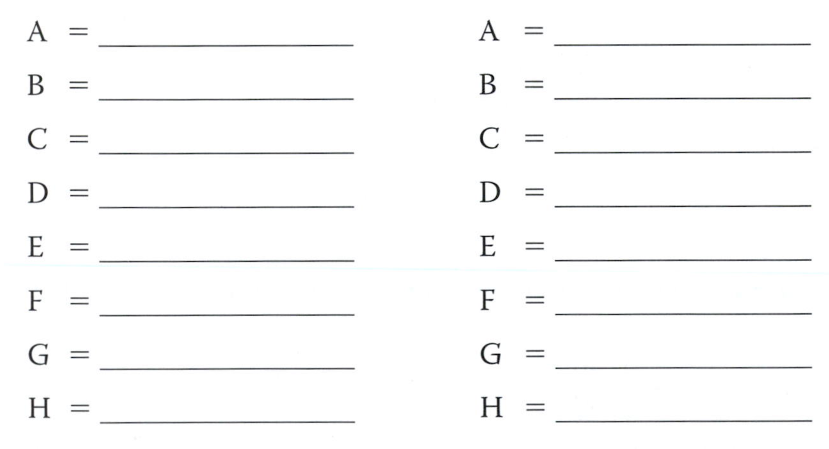

Millimeters	Centimeters
A = ________	A = ________
B = ________	B = ________
C = ________	C = ________
D = ________	D = ________
E = ________	E = ________
F = ________	F = ________
G = ________	G = ________
H = ________	H = ________

2. This piece of steel channel has a length of 22 millimeters. Express this measurement in centimeters. ______________

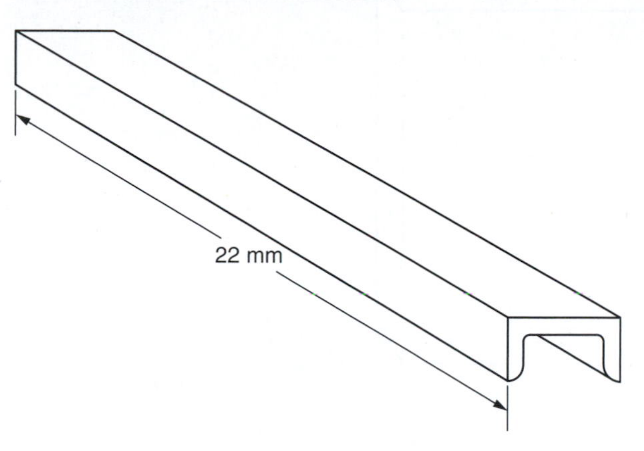

3. How many centimeters are there in 1 meter? ______________

4. A pipe with end plates is shown.

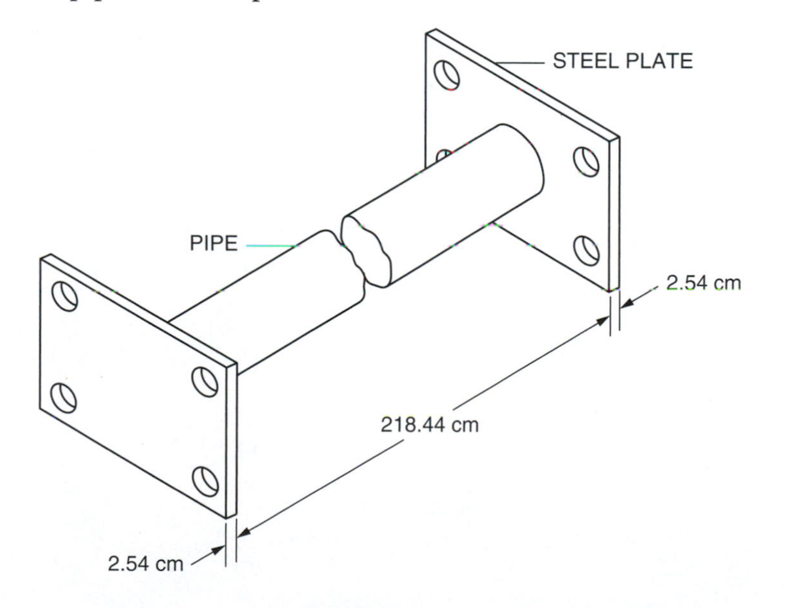

a. Find the length of the pipe section in the weldment in millimeters. a. ______________

b. Find the thickness of one end plate in millimeters. b. ______________

c. Find the overall length in meters. c. ______________

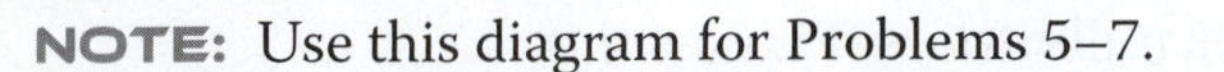

NOTE: Use this diagram for Problems 5–7.

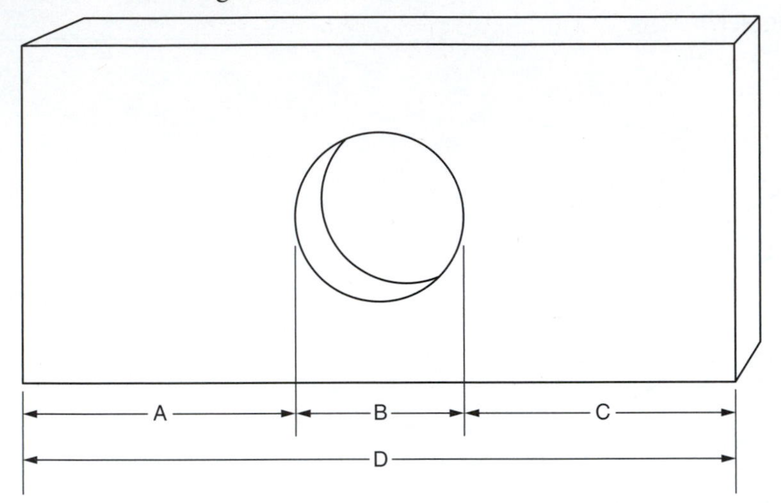

5. A = 36 cm

   B = 28 cm

   C = 384 mm

   D = ___?___ cm   D. ________________

6. A = .50 m

   B = 42 cm

   C = 530 mm

   D = ___?___ m   D. ________________

7. A = 254 mm

   B = 178 mm

   C = ___?___ mm

   D = 72.3 cm   C. ________________

8. This piece of bar stock is cut into pieces, each 7 centimeters long. How many pieces are cut? Disregard cutting waste (a.); allow .045 cm kerf width (b.).

a. ______________

b. ______________

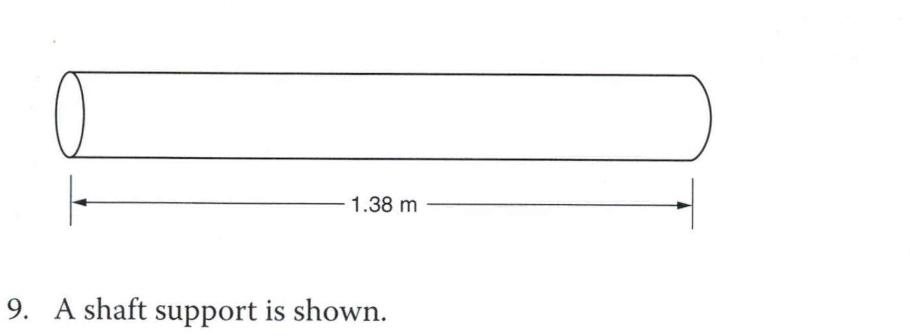

9. A shaft support is shown.

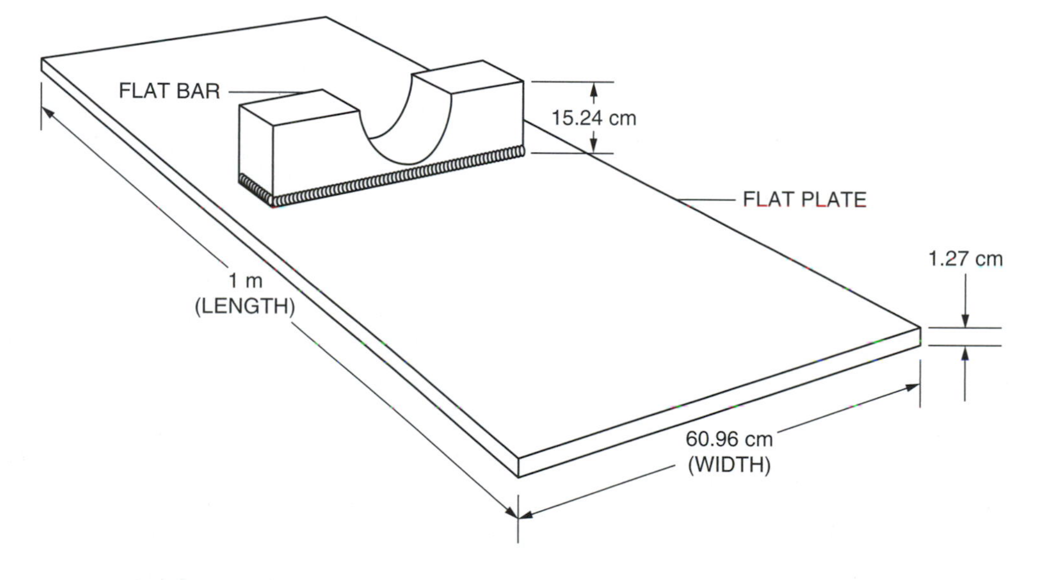

a. Find the overall height of the shaft support in centimeters. a. ______________

b. Express the length of the steel plate in millimeters. b. ______________

c. Express the width of the steel plate in centimeters. c. ______________

10. This shaft is turned on a lathe from a piece of cold-rolled round stock.

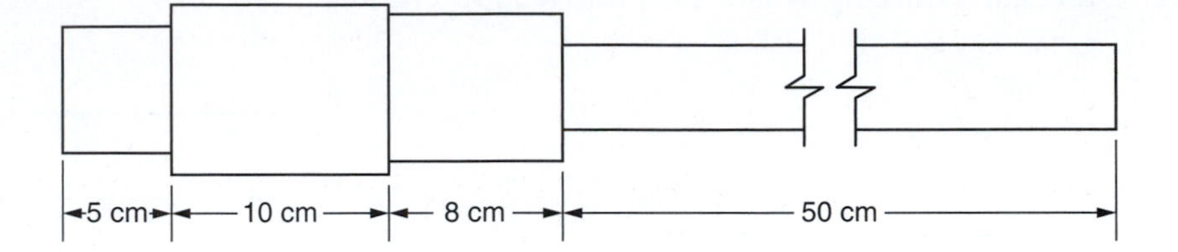

a. Find the total length in centimeters. a. ______________

b. Find the total length in meters. b. ______________

11. Nine pieces of this pipe are welded together to form a continuous length. What is the length, in meters, of the welded section? ______________

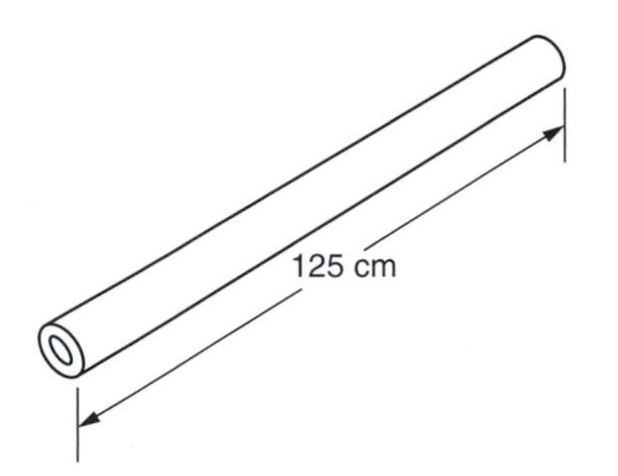

UNIT 23

English-Metric Equivalent Unit Conversions

Basic Principles

Study this table of English-metric equivalents.

English-Metric Equivalents		
1 inch (in)	=	25.4 millimeters (mm)
1 inch (in)	=	2.54 centimeters (cm)
1 foot (ft)	=	0.3048 meter (m)
1 yard (yd)	=	0.9144 meter (m)
1 mile (m)	=	1.609 kilometers (km)
1 millimeter (mm)	=	0.03937 inch (in)
1 centimeter (cm)	=	0.39370 inch (in)
1 meter (m)	=	3.28084 feet (ft)
1 meter (m)	=	1.09361 yards (yd)
1 kilometer (km)	=	0.62137 mile (mi)

When converting English to metric, or metric to English, use the table above.

EXAMPLE: Convert 2 meters into feet. 1 meter = 3.28084 feet.

Reminder: Round off the answer only after calculations are made.

$$\begin{array}{rl} 3.28084 & \text{(feet)} \\ \times \quad 2 & \\ \hline 6.56168 & \text{feet} \end{array} \rightarrow \text{rounded off} = 6.56 \text{ feet}$$

EXAMPLE: Convert 3 feet into meters. 1 foot = 0.3048 meters.

$$\begin{array}{rl} 0.3048 & \\ \times \quad 3 & \\ \hline 0.9144 & \text{meters} \end{array} \rightarrow \text{rounded off} = 0.91 \text{ meters}$$

Practical Problems

NOTE: Use this diagram for Problems 1 and 2.

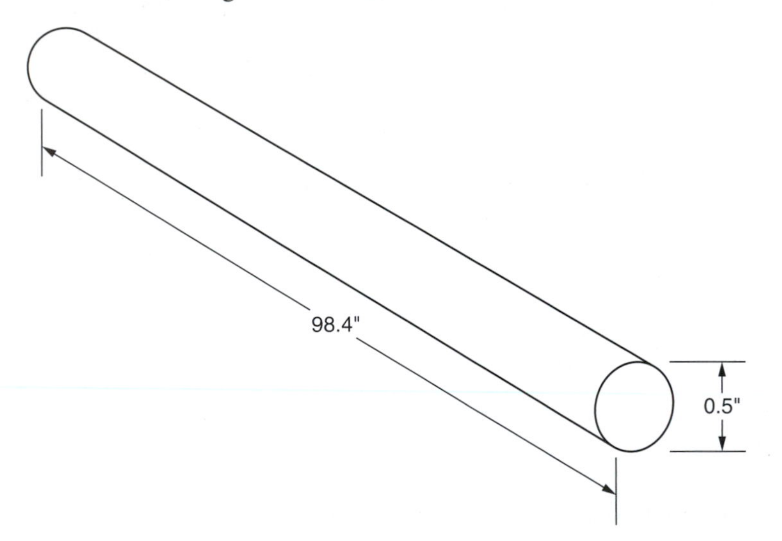

1. The round stock is 98.4″ long. Express this length in meters. Round the answer to the nearest thousandth meter. __________

2. Find the diameter of the round stock to the nearest hundredth millimeter. __________

3. This I beam is 180 cm long and 14.5 cm high. Round each answer to two decimal places.

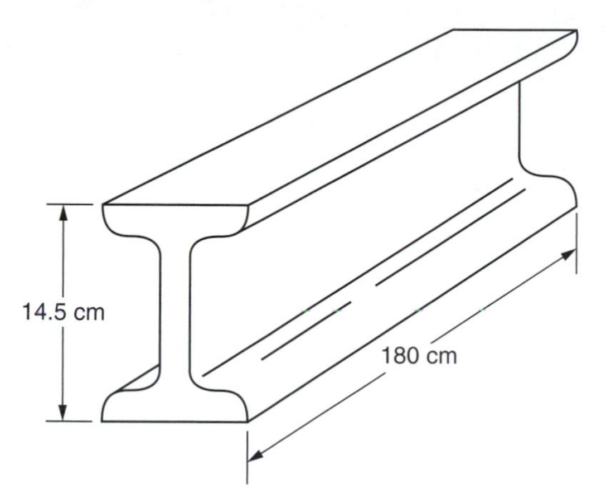

a. Express the length in inches. a. __________

b. Express the height in inches. b. __________

4. A piece of plate stock is shown.

0.75"
(THICKNESS)
254"
(LENGTH)
30.48"
(WIDTH)

a. Express the plate thickness in centimeters. a. __________

b. Express the plate width in centimeters and meters. b. _____ _____

c. Express the plate length in centimeters and meters. c. _____ _____

5. Express in meters the length and width of the following object. Express the height in millimeters.

NOTE: Convert fractions to decimals to solve Problems 5 and 6.

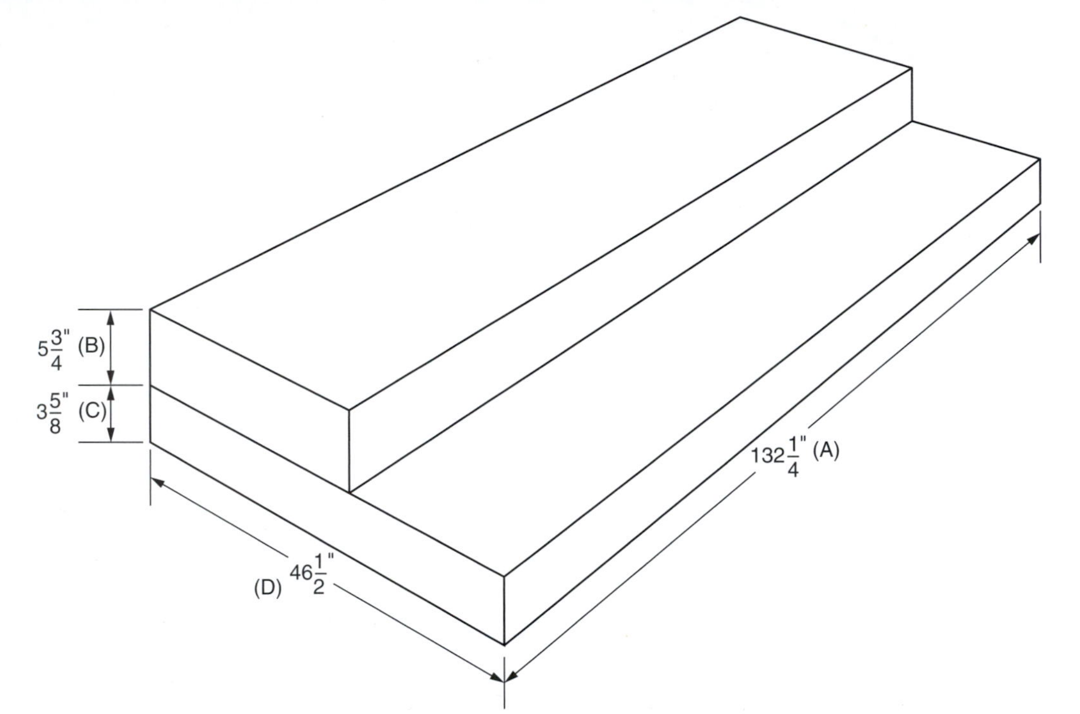

a. Dimension A a. ______________

b. Dimension B b. ______________

c. Dimension C c. ______________

d. Dimension D d. ______________

6. Express each measurement in millimeters.

a. $\frac{1}{16}$ inch — a. ______________

b. $\frac{1}{8}$ inch — b. ______________

c. $\frac{3}{16}$ inch — c. ______________

d. $\frac{1}{4}$ inch — d. ______________

e. $\frac{1}{2}$ inch — e. ______________

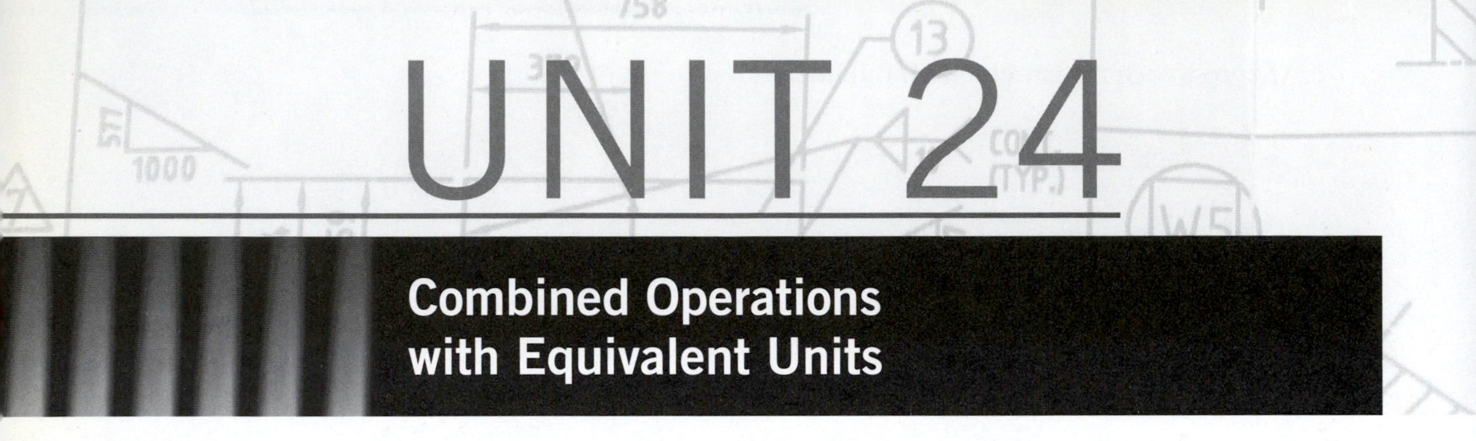

UNIT 24

Combined Operations with Equivalent Units

Basic Principles

Review the principles of operations from previous chapters and apply them to these problems.

Review the tables of equivalent units in Section 2 of the Appendix.

Practical Problems

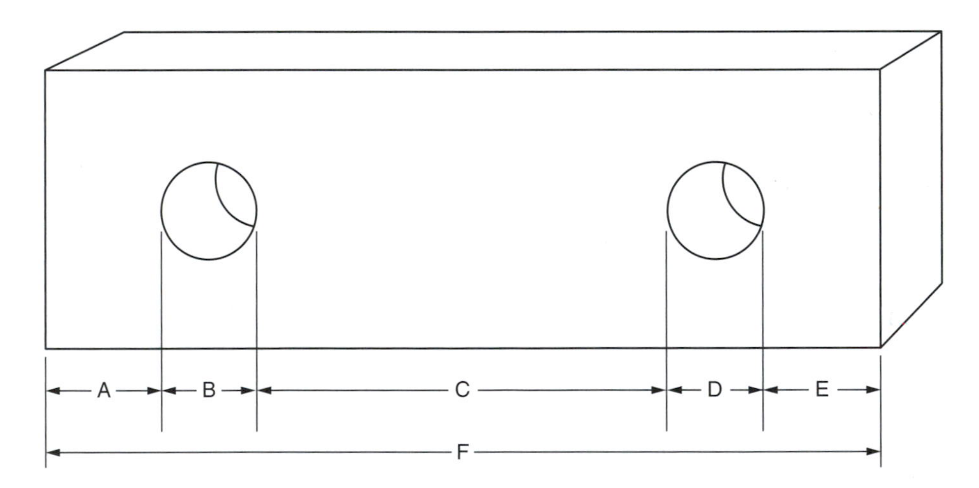

NOTE: If the above spacer block has the following dimensions, solve for the unknown in Problems 1–3.

1. A = 3.81 cm

   B = 2.54 cm

   C = 22.86 cm

   D = 2.54 cm

   E = 3.81 cm

   F = ? inches

F = ______ inches

2. $A = 1\frac{1}{4}''$

   $B = \frac{3}{4}''$

   C = ? cm

   $D = \frac{3}{4}''$

   $E = 1\frac{1}{4}''$

   F = 12″

C = ______ cm

3. A = 1.625″

   B = 22.75 mm

   C = 3.28 cm

   D = 22.75 mm

   E = 1.375″

   F = ? m

F = ______ m

4. This drawing shows a welded pipe support.

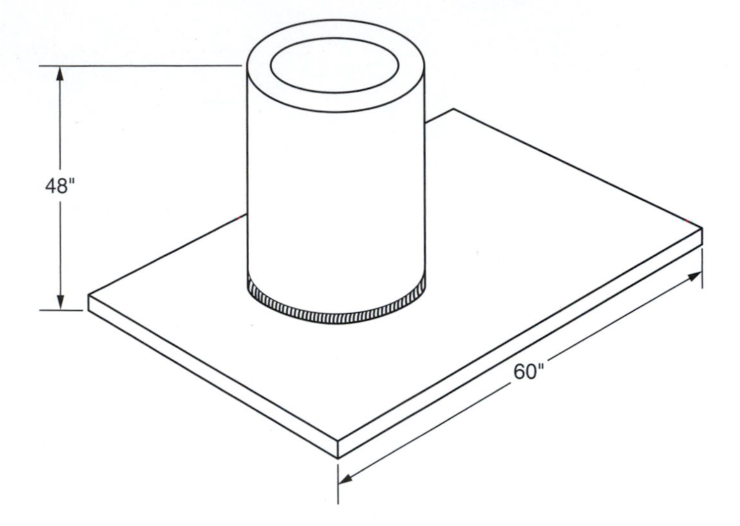

a. Express the height in meters. a. ____________

b. Express the width in meters, centimeters, and feet. b. ____________

____

____

5. This steel gusset is a right angle (90°) triangle. Round each answer to two decimal places.

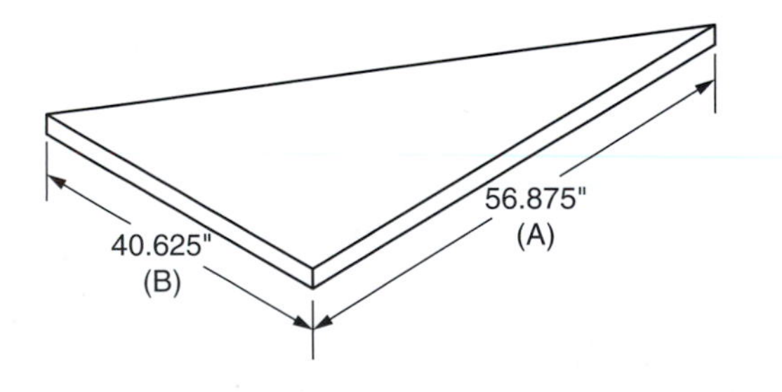

a. Express side A in centimeters. a. ____________

b. Express side B in centimeters and millimeters. b. ____________

____

6. A pipe bracket is shown. Round all answers to three decimal places.

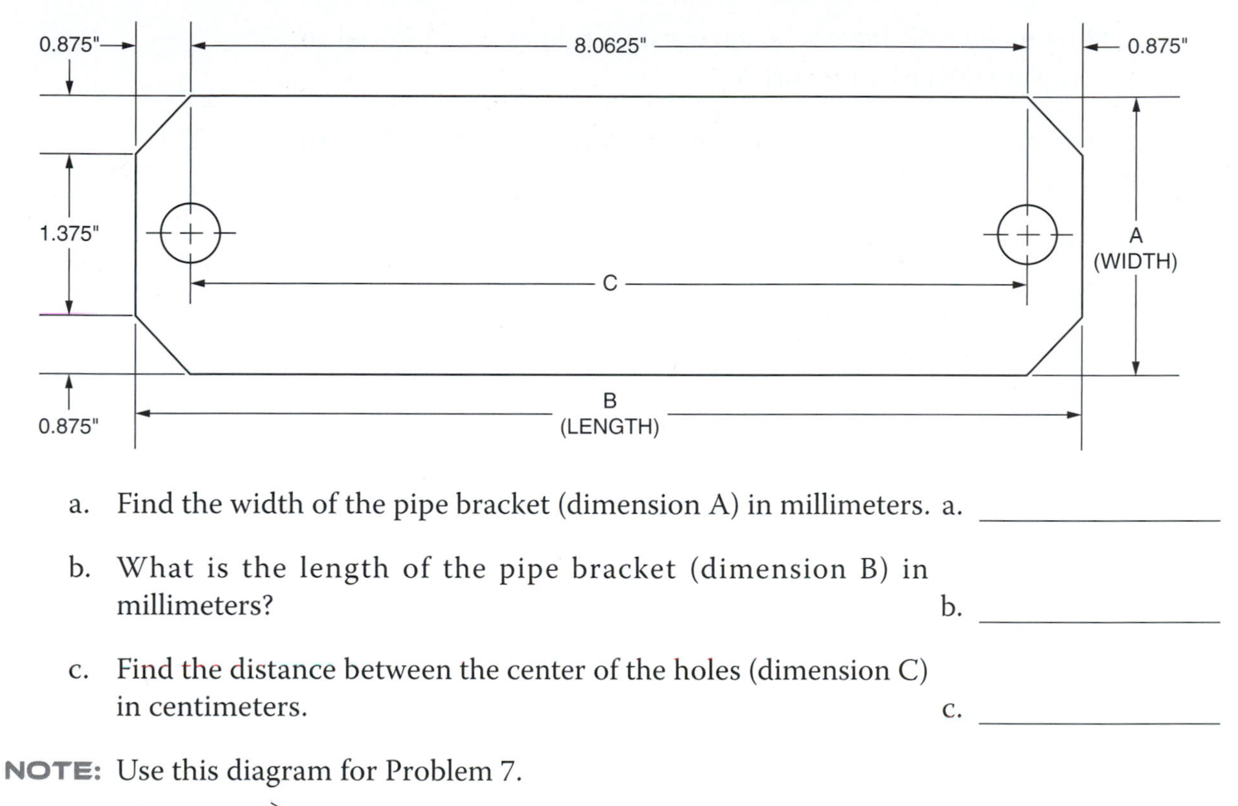

a. Find the width of the pipe bracket (dimension A) in millimeters. a. ______________

b. What is the length of the pipe bracket (dimension B) in millimeters? b. ______________

c. Find the distance between the center of the holes (dimension C) in centimeters. c. ______________

NOTE: Use this diagram for Problem 7.

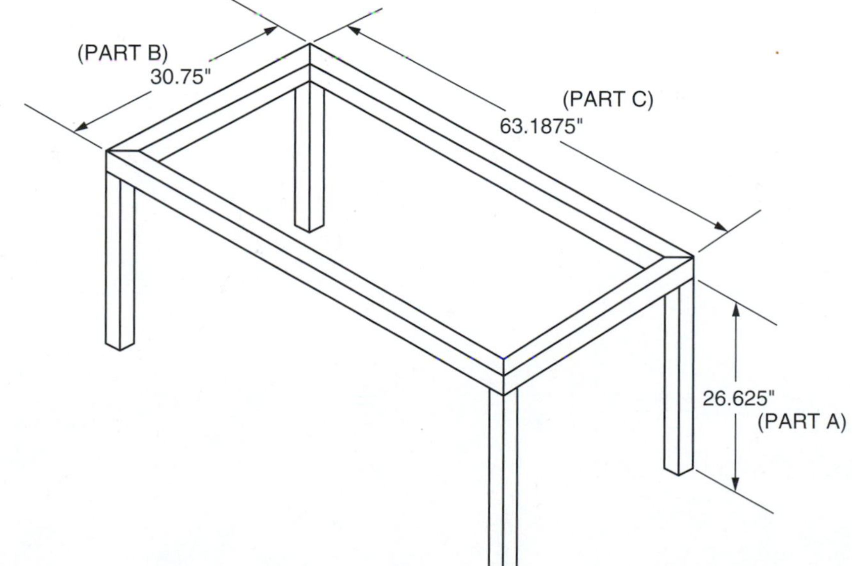

7. A welder makes 20 of these table frames.

   a. How many centimeters of square steel tubing are required to complete the order for Part A? a. ______________

   b. How many centimeters of square steel tubing are required to complete the order for Part B? b. ______________

   c. How many meters of square steel tubing are required to complete the order for Part C? c. ______________

SECTION 6

COMPUTING GEOMETRIC MEASURE AND SHAPES

UNIT 25

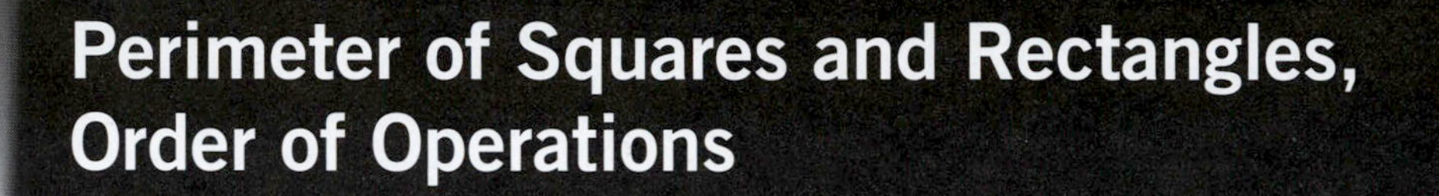

Perimeter of Squares and Rectangles, Order of Operations

Basic Principles

Definitions

Perimeter: The distance around a figure; the sum of its sides.

Square: A four-sided figure, as shown below. All four sides are of equal length, and all four angles are 90°.

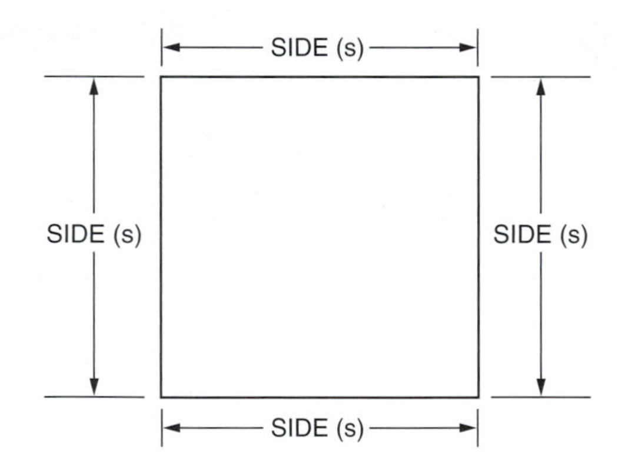

Rectangle: A four-sided figure, as shown below. The lengths are equal only to each other and the widths are equal only to each other. All four angles are 90°.

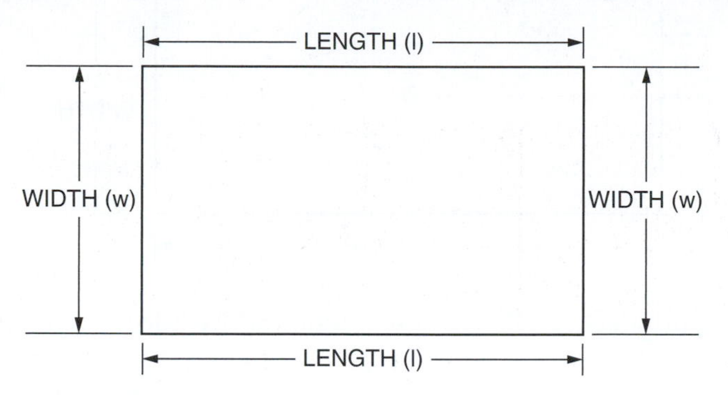

Exponents are used to indicate the math instruction of multiplying a number times itself.

NOTE: Length is designated on drawings with the letters L, l, or ℓ.

EXAMPLES:

a. $4^2 = 4 \times 4$
$4^2 = 16$

b. $9^2 = 9 \times 9$
$9^2 = 81$

c. $3^3 = 3 \times 3 \times 3$
$3^3 = 27$

d. $(3 + 2)^2 = (5)^2 = 5 \times 5$
$(3 + 2)^2 = 25$

Formulas are used to calculate the perimeter, area, or volume of geometric shapes. A formula is a set of math instructions that solve a specific problem.

EXAMPLE 1: What is the perimeter of square A? All four sides are 8″ in length.

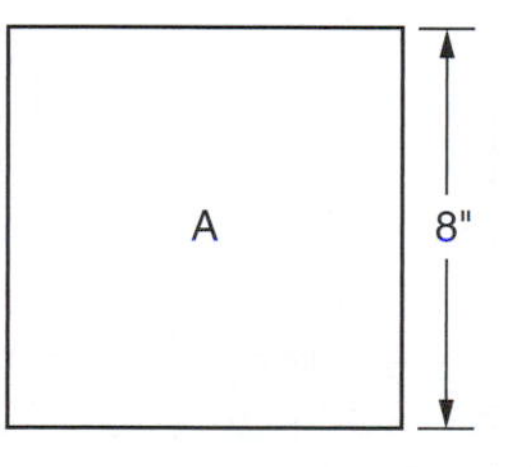

Since the distance around a square can be calculated by multiplying the side length by four, the formula used is:

$$P = 4s$$

SOLUTION:

$P = 4s$
$P = 4 \times 8$
$P = 32''$

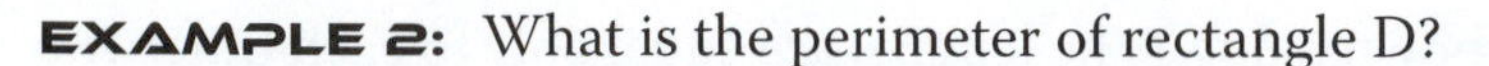

EXAMPLE 2: What is the perimeter of rectangle D?

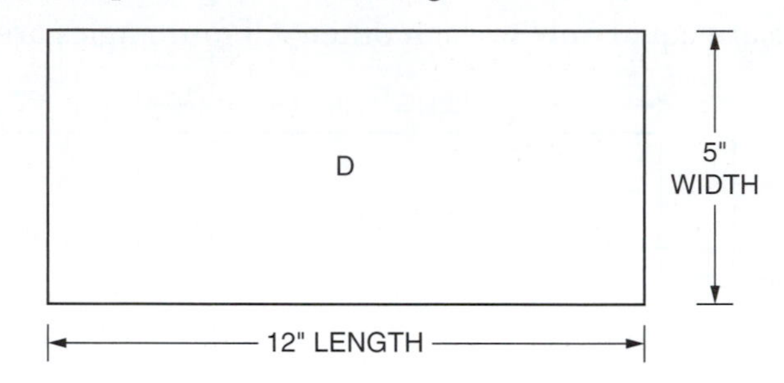

Since the distance around a rectangle can be calculated by adding 2 lengths and 2 widths, the formula used is:

$$P = 2\ell + 2w$$

SOLUTION:

$$\begin{aligned} P &= 2\ell + 2w \\ &= (2 \times 12) + (2 \times 5) \\ &= 24 + 10 \\ P &= 34'' \end{aligned}$$

Order of Operations

To calculate mathematical problems, we follow the steps in what is known as the "order of operations."

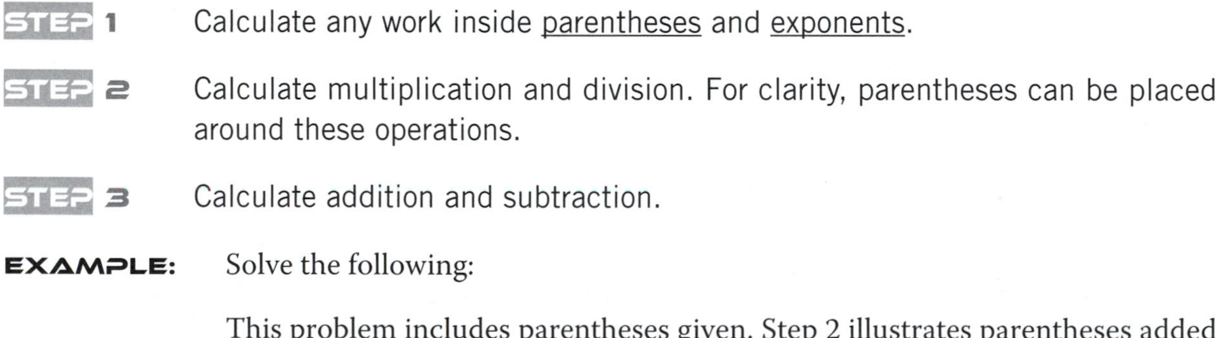

STEP 1 Calculate any work inside parentheses and exponents.

STEP 2 Calculate multiplication and division. For clarity, parentheses can be placed around these operations.

STEP 3 Calculate addition and subtraction.

EXAMPLE: Solve the following:

This problem includes parentheses given. Step 2 illustrates parentheses added for clarifying the order of operations.

$$3^2 + 8 \times (5 - 3) - 4 \div 2 = ?$$

SOLUTION:

STEP 1 $3^2 + 8 \times (5 - 3) - 4 \div 2 =$ → $\mathbf{9} + 8 \times \mathbf{2} - 4 \div 2 =$

Exponents and math inside parentheses are calculated first.

STEP 2 $9 + (8 \times 2) - (4 \div 2) =$ → $9 + \mathbf{16} - \mathbf{2} =$

× and ÷ are calculated next.

+ and − are calculated next.

STEP 3 $25 - 2 =$

$25 - 2 = 23$

ANSWER: $3^2 + 8 \times (5 - 3) - 4 \div 2 = 23$

Practical Problems

1. The measure of one side of square plates is given. Calculate the perimeter of each plate.

   a. $1\frac{3}{4}''$ __________

   b. 16 cm __________

   c. 193.675 mm __________

   d. 9.5″ __________

   e. .78 m __________

2. Find the perimeter of the following rectangles:

   a. length = 17 cm; width = 8 cm __________

   b. length = 92″; width = 43″ __________

   c. ℓ = 22 mm; w = 17.5 mm __________

UNIT 26

Area of Squares and Rectangles

Basic Principles

The formulas for calculating the area (A) of squares and rectangles are described below:

Formula for the Area of Squares

$$A = s^2$$

Reminder

The exponent (^{2}), when used in the formula, instructs multiplication of side $\times$ side.

Area: Two dimensional space—length and width

The following illustrates the development of 1 square foot into square inches.

DEVELOPMENT OF SQUARE INCHES IN 1 SQUARE FOOT

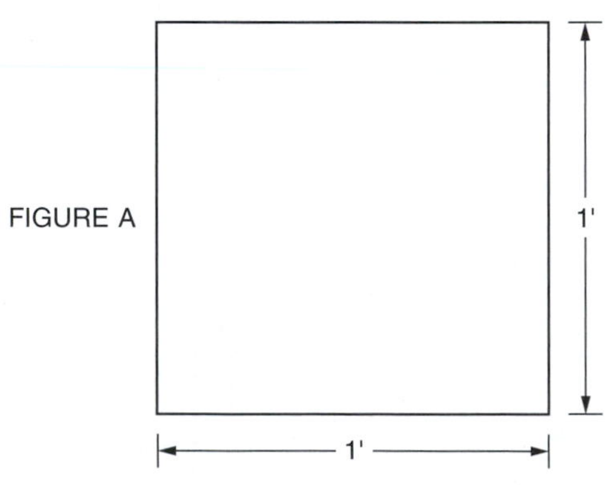

SOLUTION: conversion of 1 square foot into square inches.

$$A = s^2$$
$$= 12'' \times 12''$$
$$= 144 \text{ in}^2$$

ANSWER: 144 in^2

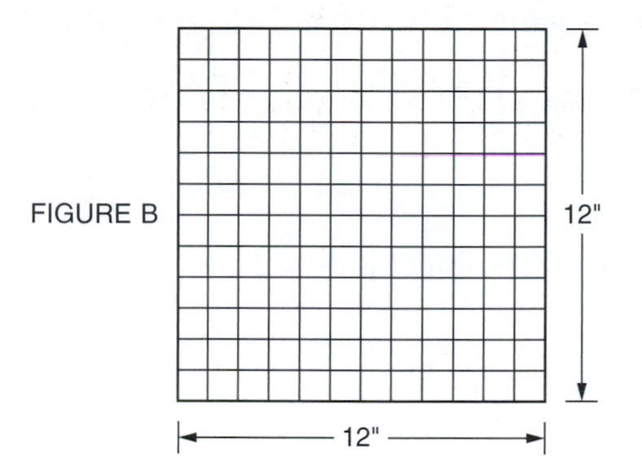

FIGURE B

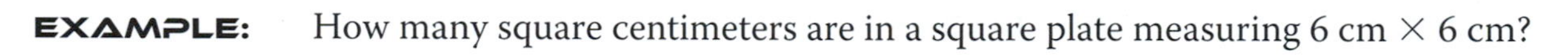

EXAMPLE: How many square centimeters are in a square plate measuring 6 cm × 6 cm?

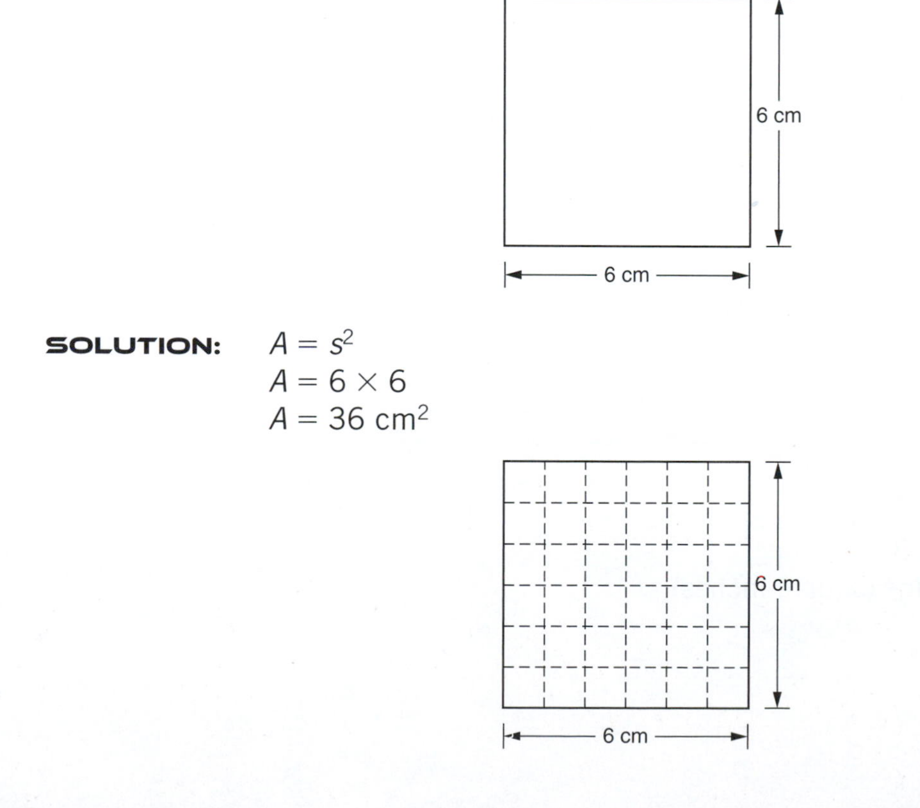

SOLUTION:

$$A = s^2$$
$$A = 6 \times 6$$
$$A = 36 \text{ cm}^2$$

The answers are written using the exponent (^{2}) again. However, in this case, the exponent is used to show that the object in the answer, square centimeters, has

a. 2 dimensions: length and width; and

b. the shape of a square.

Study the work in the solution again. Notice that the exponent has two different uses:

Its first use is as a math instruction.

$$s^2 = \text{side} \times \text{side}$$

Its second use is to describe what an object looks like.

cm^2: a square centimeter, 2 dimensional

Formula for the Area of Rectangles

$$A = l \times w$$

To find the area of a rectangle, multiply the length × the width.

EXAMPLE: How many square inches are in a rectangular plate measuring 8″ × 3″?

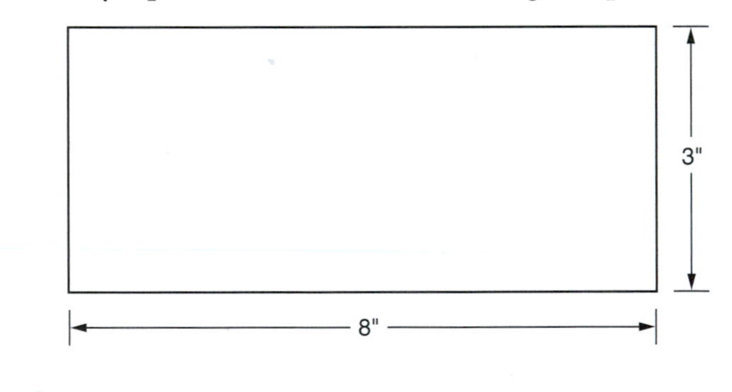

SOLUTION:

$$\begin{aligned} A &= lw \\ &= 8 \times 3 \\ &= 24 \text{ in}^2 \text{ (square inches)} \end{aligned}$$

ANSWER: 24 in^2

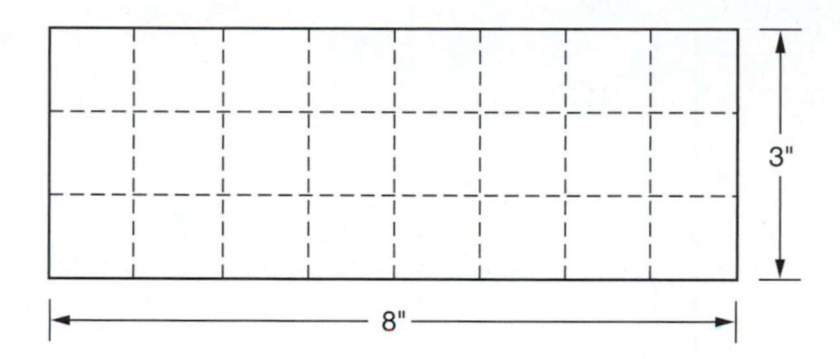

Practical Problems

These squares are made from 16-gauge sheet metal. Find the area of each square.

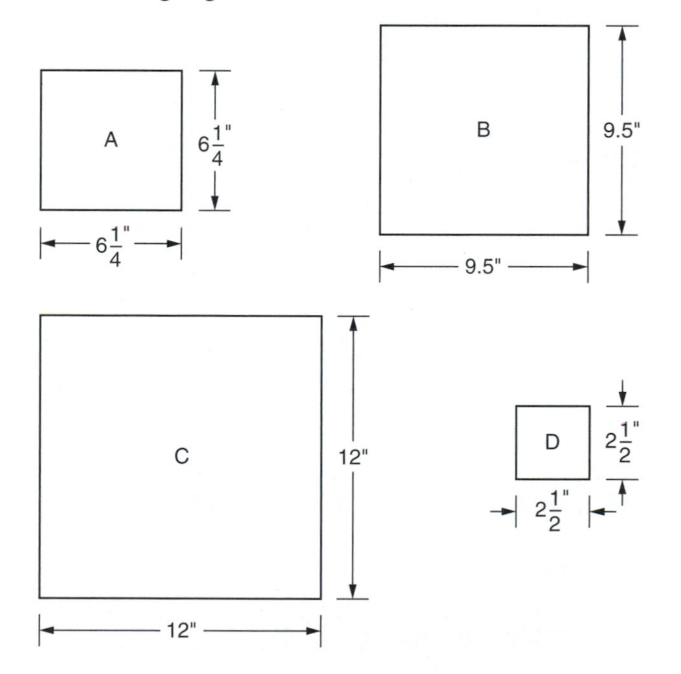

1. Square A ________
2. Square B ________
3. Square C ________
4. Square D ________

5. How many square inches are in 1 square foot? _______________

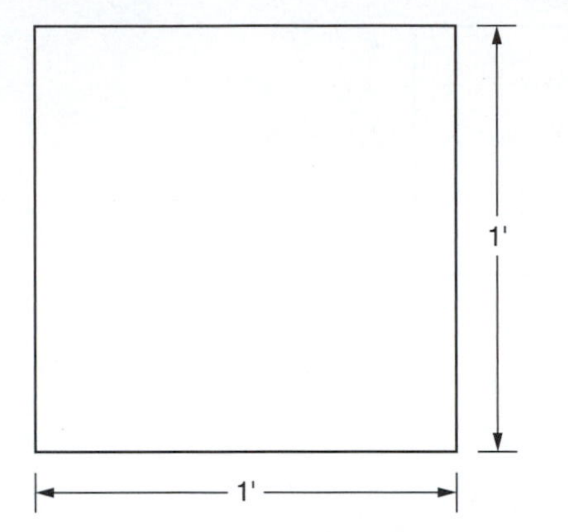

NOTE: Use this diagram for Problems 6 and 7.

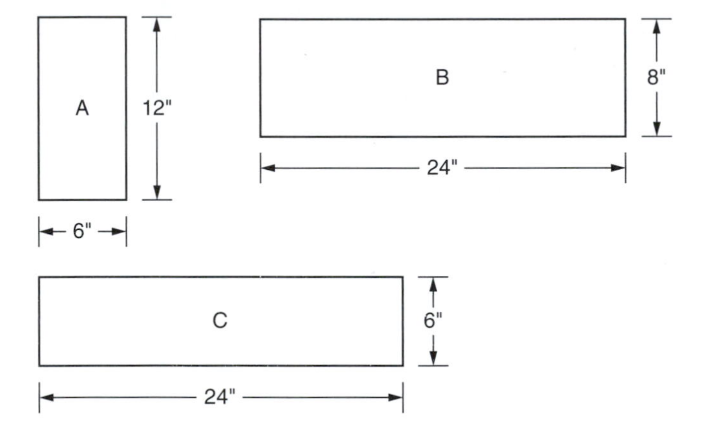

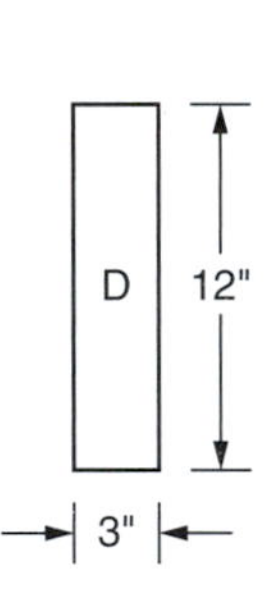

6. The four pieces of sheet metal are cut for a welding job.

   a. Find the area of rectangle A in square inches. a. _______________

   b. Find the area of rectangle B in square inches. b. _______________

   c. Find the area of rectangle C in square inches. c. _______________

   d. Find the area of rectangle D in square inches. d. _______________

   e. What is the total area of the pieces in square inches? e. _______________

   f. Express the total area in square feet. Round the answer to two decimal places. f. _______________

7. Which of the pieces has an area of 1 square foot? __________

8. A rectangular tank is made from plates with the dimensions shown. Find the total area of plate needed to complete the tank in square inches. __________

   How many square feet of plate is needed? __________

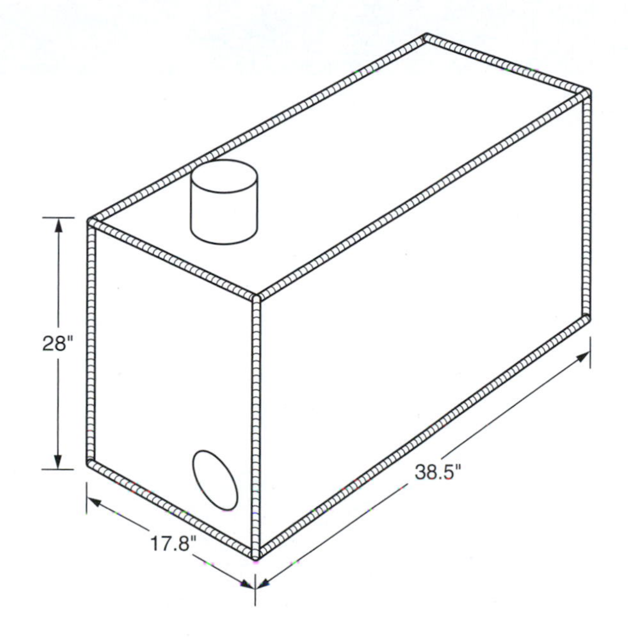

UNIT 27

Area of Triangles and Trapezoids

Basic Principles

Triangles/Rectangles

Triangle ABC is half of a rectangle (see illustrations.) It is a three-sided figure containing three angles totaling 180°.

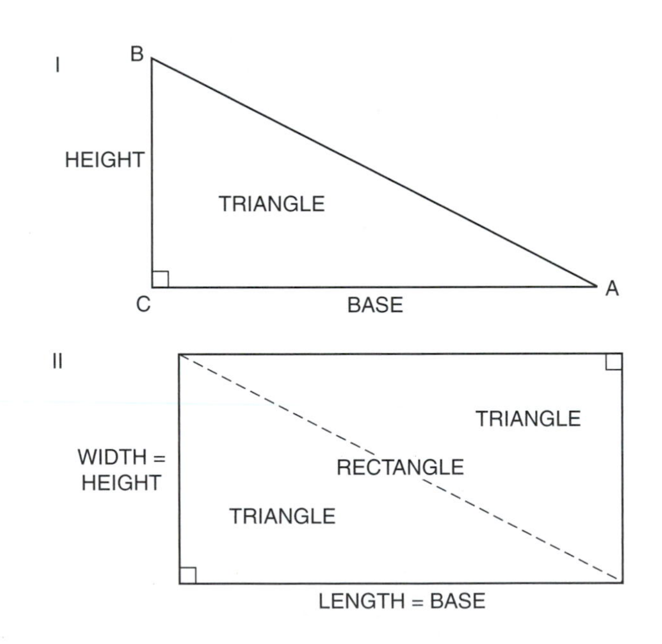

NOTE: The length of the rectangle is the same label as the base of the triangle.

The width of the rectangle is the same label as the height of the triangle.

Formula

Because a triangle is ½ of a rectangle, the formula for determining the area (A) of a triangle is:

$$A = \frac{1}{2}(\text{base} \times \text{height})$$

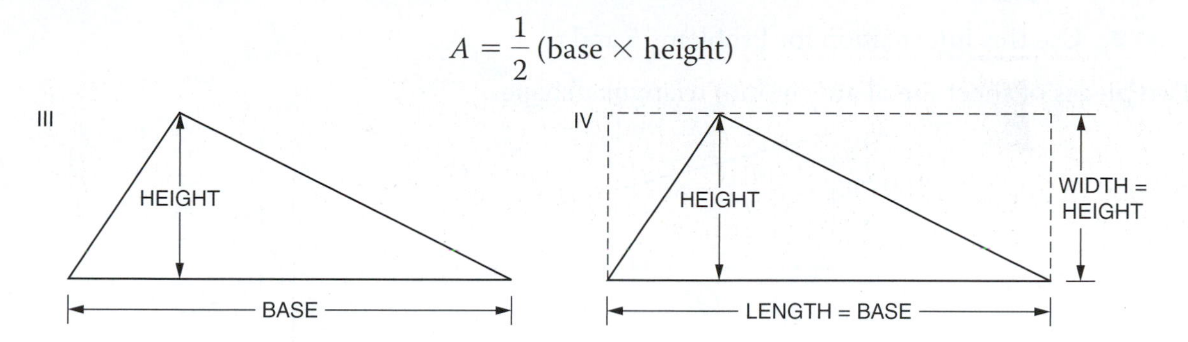

Practical Problems

NOTE: Use this information for Problems 1–4.

These four triangular shapes are cut from sheet metal. What is the area of each piece in square inches?

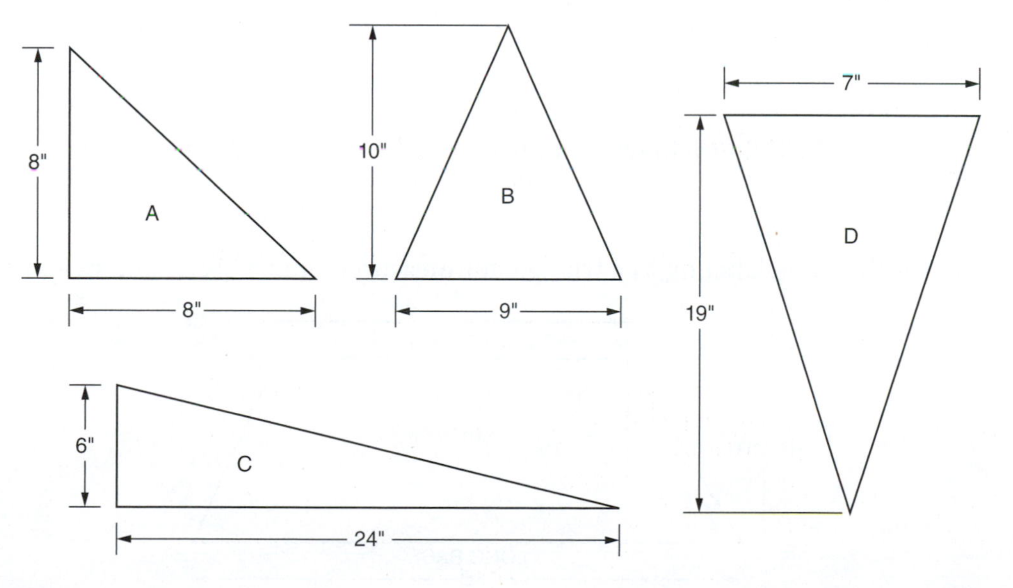

1. Triangle A ____________
2. Triangle B ____________

3. Triangle C ______________

4. Triangle D ______________

NOTE: Use this information for Problems 5 and 6.

Two pieces of sheet metal are cut into triangular shapes.

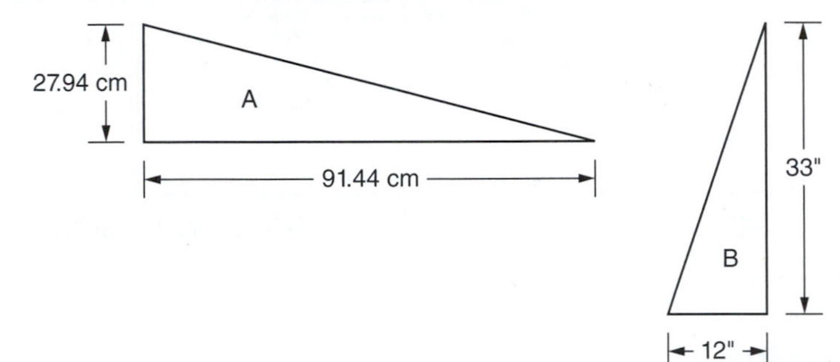

5. Find, in square centimeters, the area of triangle A. ______________

6. Find, in square inches, the area of triangle B. ______________

Trapezoids

Definition

A trapezoid is a four-sided figure in which only two of the sides are parallel.

Labeling

A trapezoid uses the same labeling as a triangle for measurements: base(s) and height.

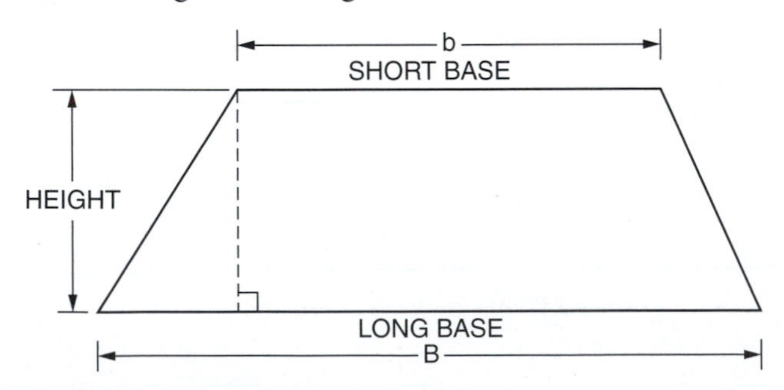

NOTE: The formula for determining the area of a trapezoid is based on the formula for the area of a rectangle: $A = \ell w$. The trapezoid has two bases (lengths), so we need to find the average base.

To find the average base, add both bases together and divide by 2.

 $\frac{B + b}{2}$ = average base

STEP 2 Multiply average base times the height.

Formula

$$A = \left(\frac{B + b}{2}\right)h$$

 Determine the area (A) of the following trapezoid.

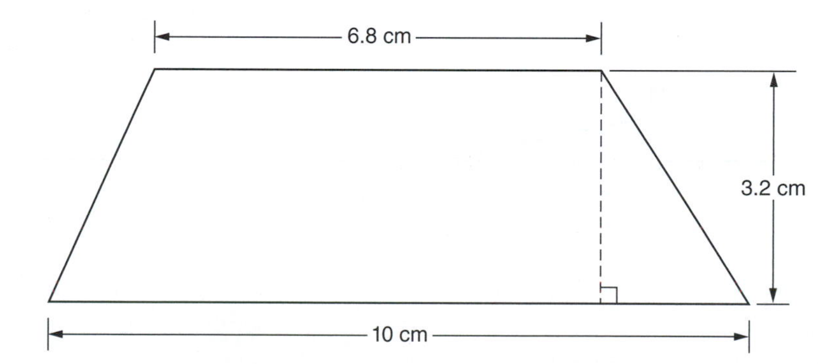

SOLUTION: $A = \left(\frac{B + b}{2}\right)h$

$$\left(\frac{10 + 6.8}{2}\right)3.2$$

$$\left(\frac{16.8}{2}\right)3.2$$

$$(8.4)3.2 = \begin{array}{r} 3.2 \\ \times 8.4 \\ \hline 128 \\ 256 \\ \hline 26.88 \end{array}$$

ANSWER: $A = 26.88\ \text{cm}^2$

NOTE: Use this information for Problems 7–10.

These four support gussets are cut from ¼-inch plate. What is the area of each piece in square inches?

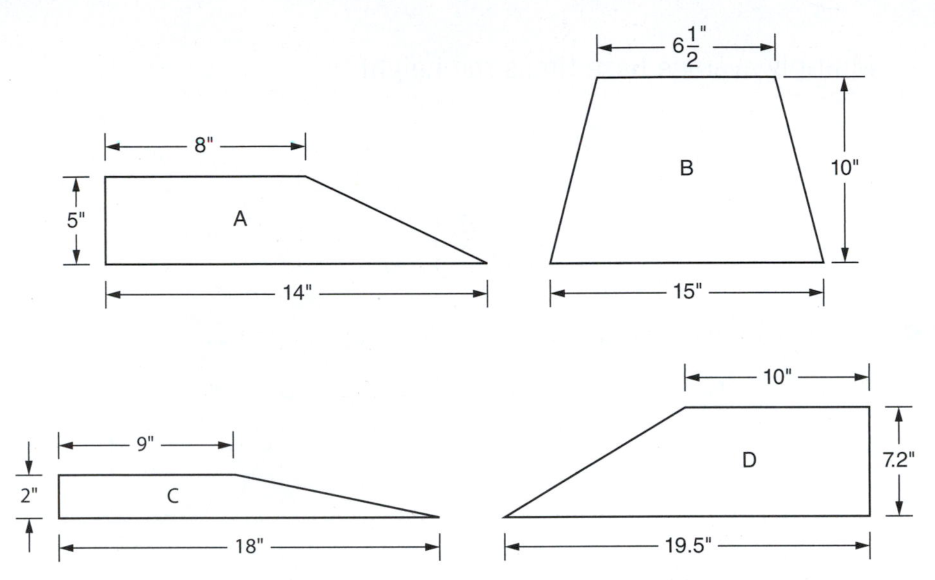

7. Gusset A __________

8. Gusset B __________

9. Gusset C __________

10. Gusset D __________

11. One-hundred-twenty support gussets are cut as shown. Find, in square feet, the total area of steel plate needed for the complete order. __________

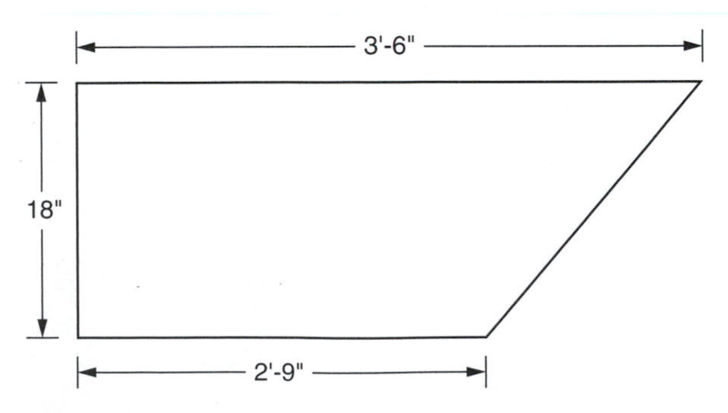

12. A welded steel bin is made from plates with these dimensions. Find, in square centimeters, the amount of plate used to complete the bin. ______________

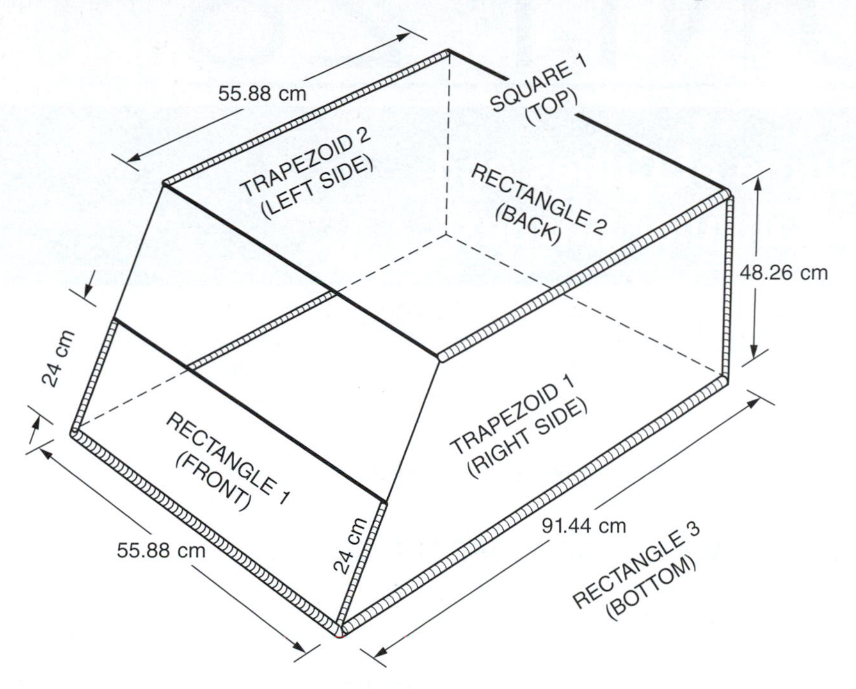

HINT: The total area equals the sum of the areas of the rectangles, the trapezoids, and the square.

UNIT 28

Volume of Cubes and Rectangular Shapes

Basic Principles

Study this table of equivalent units of volume measure for solids.

ENGLISH VOLUME MEASURE FOR SOLIDS

1 cubic yard (cu yd) = 27 cubic feet (cu ft)
1 cubic foot (cu ft) = 1,728 cubic inches (cu in)

NOTE: Use the following information for Problems 1–5.

The amount of space occupied in a three-dimensional figure is called the volume. Volume is also the number of cubic units equal in measure to the space in that figure.

The formula for the volume of a cube is:

$$\text{Volume} = \text{side} \times \text{side} \times \text{side}$$

or

$V = s^3$ The exponent (^{3}) describes the math procedure.

EXAMPLE: What is the volume of the illustrated cube?

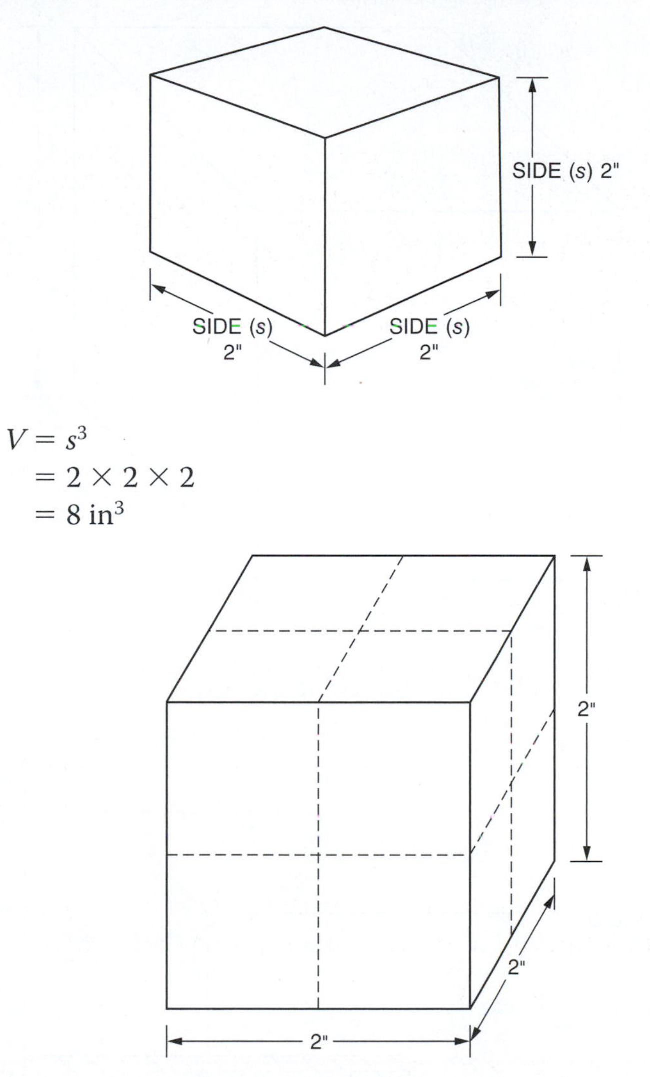

$$V = s^3$$
$$= 2 \times 2 \times 2$$
$$= 8 \text{ in}^3$$

The volume of the cube is 8 in^3. It contains 8 cubic inches of space or material.

The exponent in the answer (^{3}) shows that the object, cubic inches, is three-dimensional: it has length, width, and height (or depth).

Volume: three-dimensional space with length, width, and height.

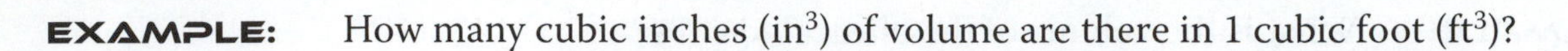

EXAMPLE: How many cubic inches (in^3) of volume are there in 1 cubic foot (ft^3)?

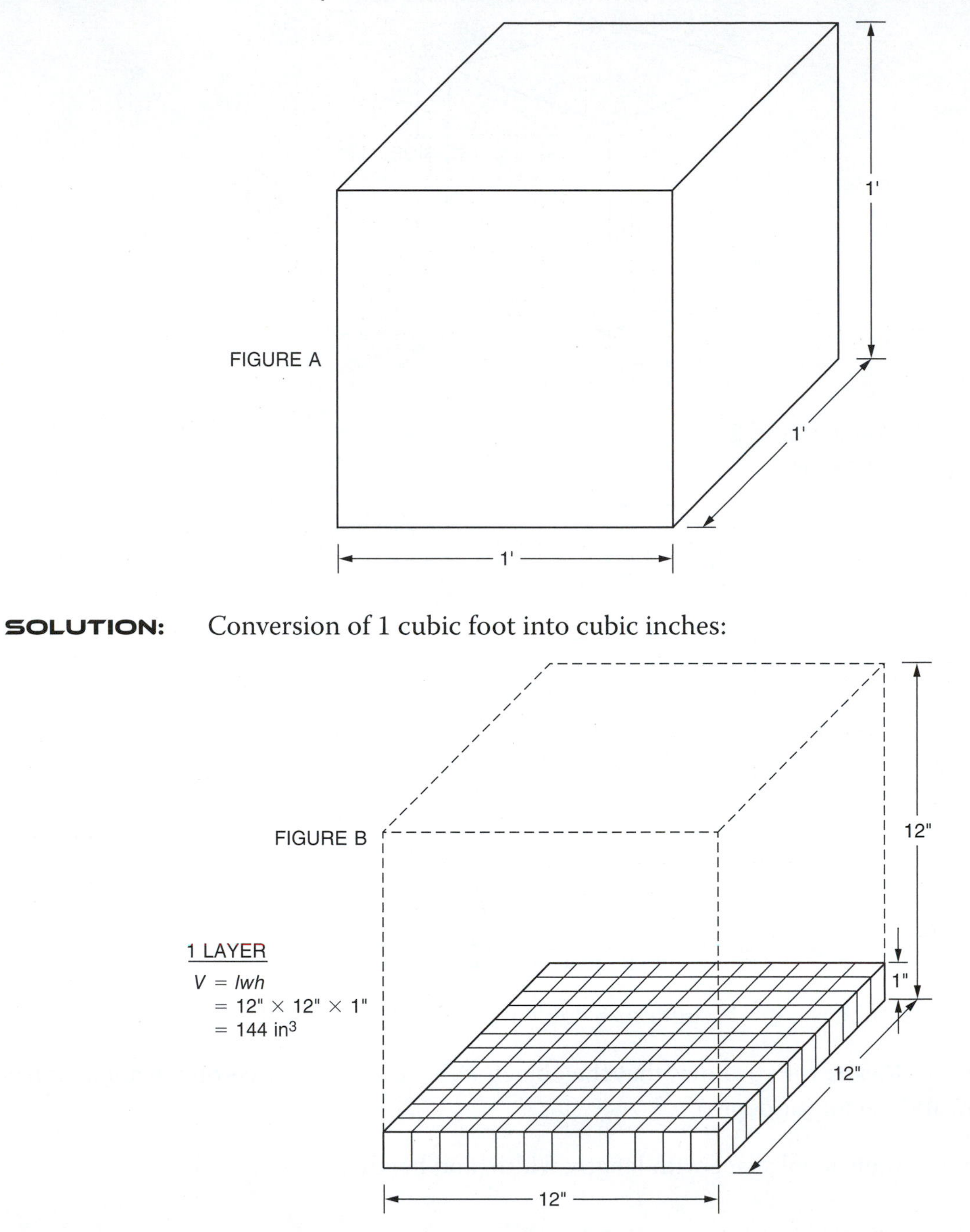

SOLUTION: Conversion of 1 cubic foot into cubic inches:

One layer has 144 in^3 of volume.

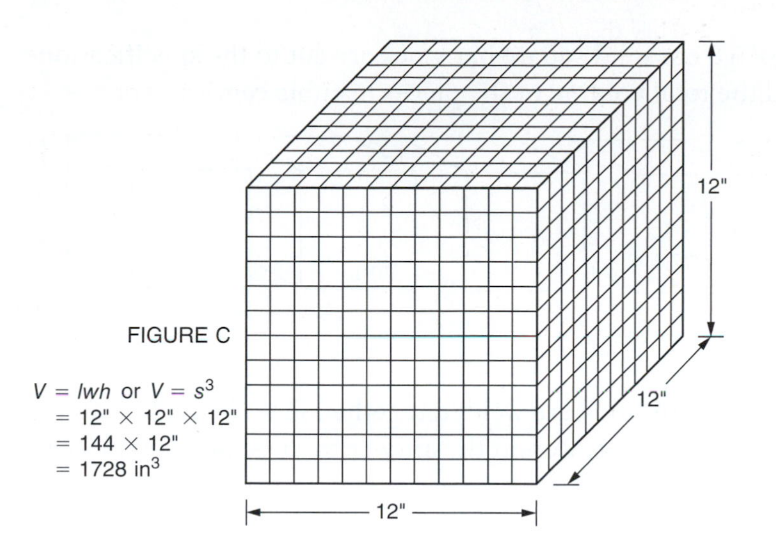

There are 12 full layers in a cubic foot.

ANSWER: Total cubic inches in 1 cubic foot (ft^3) = 1728 in^3

Practical Problems

1. A solid cube of steel is cut to these dimensions. Find the volume of the cube in cubic inches. Find the volume in cubic feet. ______ ______

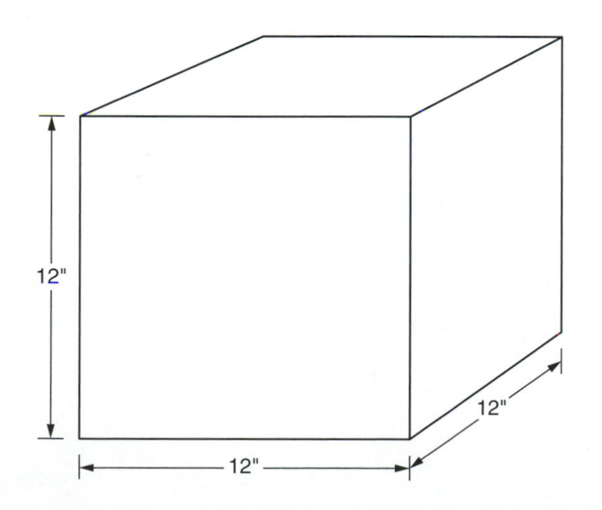

2. Five pieces of 5.8 cm solid square bar stock are cut to the specifications shown. Find the total volume of the pieces in cubic centimeters. ______________

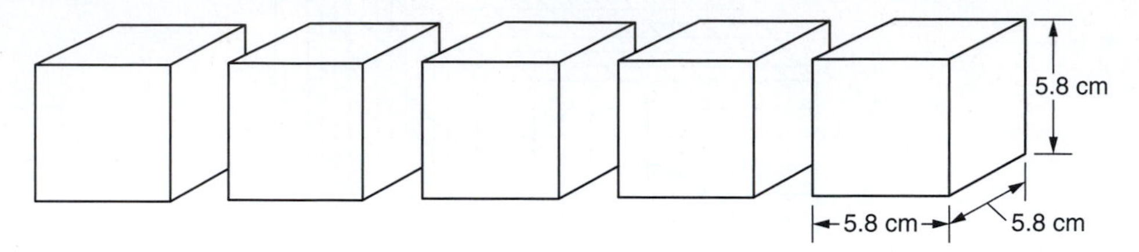

3. Two pieces of square stock are welded together. Find, in cubic feet, the total volume of the pieces. Round the answer to three decimal places. ______________

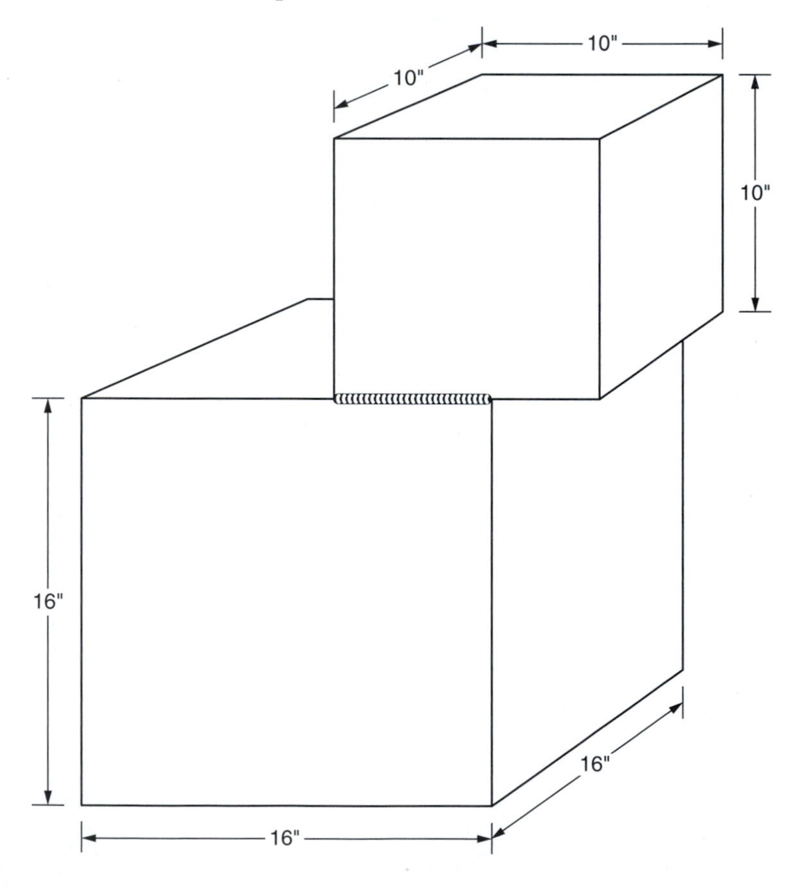

4. Sheet metal is bent to form this cube. What is the volume of the completed cube in cubic inches? ________________

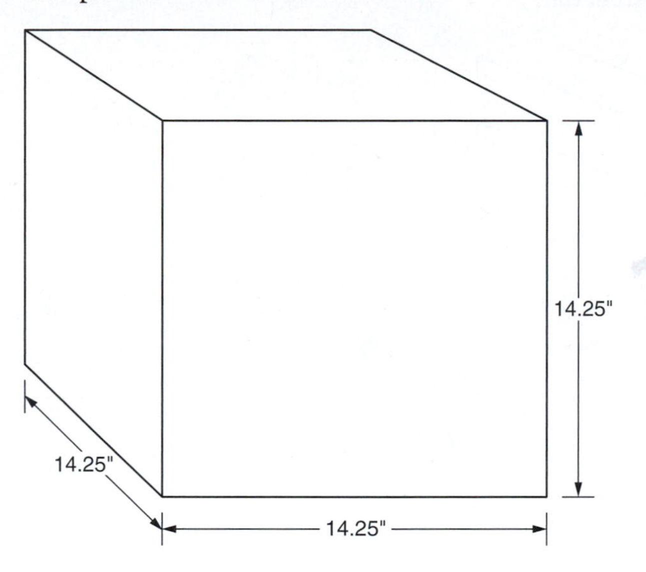

The volume of a rectangularly shaped object is calculated with the formula:

or

$$V = \ell wh$$

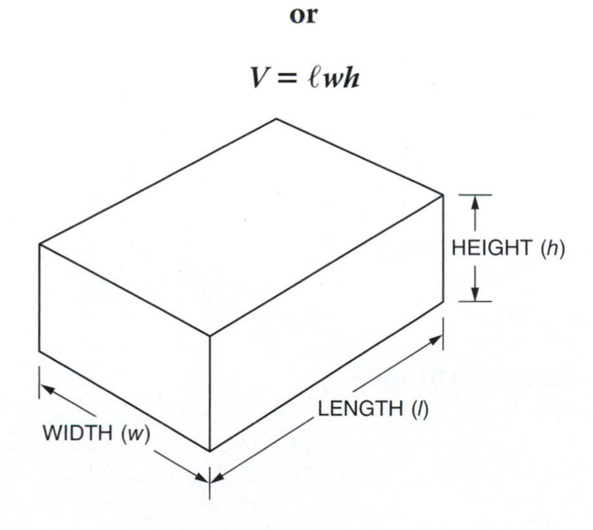

NOTE: Use this information for Problems 5–7.

Find, in cubic inches, the volume of each steel bar.

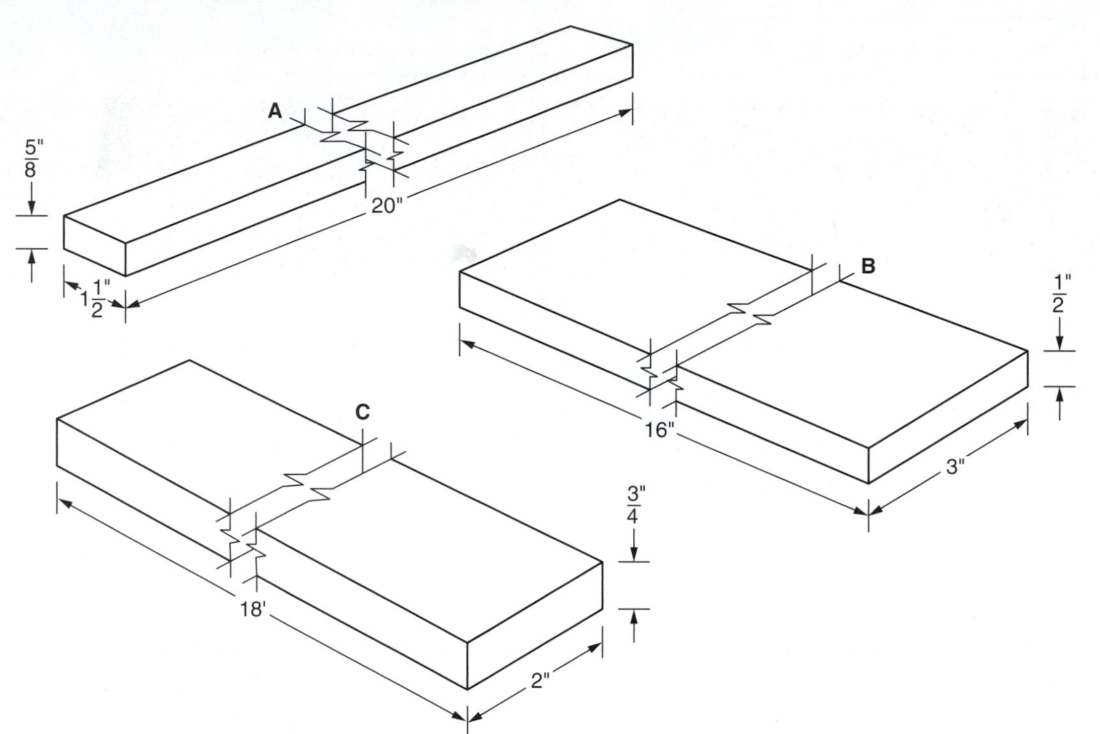

5. Steel bar A ____________

6. Steel bar B ____________

7. Steel bar C ____________

Find the volume of each rectangular solid.

8. $\ell = 12$ in; $w = 8$ in; $h = 10$ in ____________

9. $\ell = 0.84$ m; $w = 0.46$ m; $h = 0.91$ m ____________

UNIT 29

Volume of Rectangular Containers

Basic Principles

Reminder

The formula for determining the volume of rectangularly shaped objects is:

$$V = \ell wh$$

In measuring the holding capacity of rectangular tanks and containers, always use inside dimensions. If outside measurements are given, the wall thicknesses are subtracted as a first step.

EXAMPLES: a. Find the volume, in cubic inches, of a tank with the following inside dimensions:

Length = 3′
Width = 18″
Height = 26″

SOLUTION:

$$\begin{aligned} V &= \ell wh \\ &= 3' \times 18'' \times 26'' \\ &= 36'' \times 18'' \times 26'' \\ &= 16{,}848 \text{ in}^3 \end{aligned}$$

b. How many gallons will the tank hold?

NOTE: There are 231 in^3 in one gallon.

SOLUTION: Divide 231 into the volume found.

$$\frac{V}{231} = \frac{16{,}848}{231} = 72.935$$

ANSWER: 72.94 gallons

Practical Problems

Find the volume, in gallons, of each rectangular welded tank. These are inside dimensions. Round each answer to three decimal places.

1. $\ell = 9.875$ in; $w = 6.1875$ in; $h = 24.125$ in __________

2. $\ell = 12\frac{3}{4}$ in; $w = 14\frac{7}{8}$ in; $h = 36\frac{1}{4}$ in __________

3. $\ell = 36$ in; $w = 18$ in; $h = 48$ in __________

4. $\ell = 23.5$ in; $w = 23.5$ in; $h = 34.5$ in __________

5. The dimensions on this box are inside dimensions. Find the number of cubic inches of volume in the box. __________

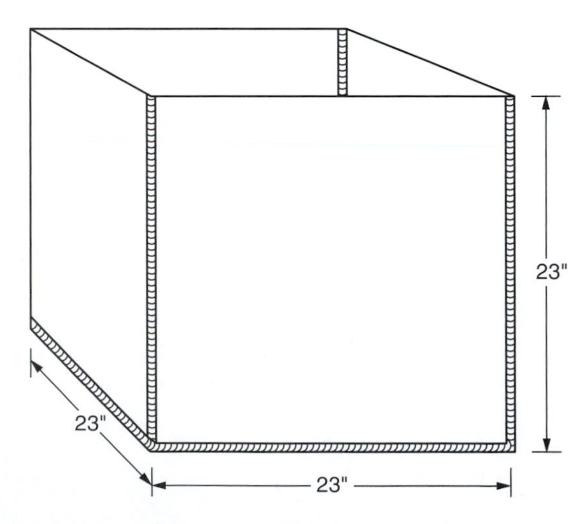

Use the following information for metric volume.

NOTE: **There are 1,000 cm^3 (cubic centimeters) in one liter. To find the number of liters a tank or container can hold, divide the volume (cm^3) by 1,000.**

EXAMPLE: A tank measuring 50 cm by 32 cm by 32 cm is built. Determine the volume in cubic centimeters, and find the number of liters the tank can hold.

SOLUTION:

STEP 1

$$\begin{aligned} V &= \ell wh \\ &= 50 \text{ cm} \times 32 \text{ cm} \times 32 \text{ cm} \\ &= 51{,}200 \text{ cm}^3 \end{aligned}$$

STEP 2

$$\frac{51{,}200}{1000} = 51.2$$

ANSWER: The tank can hold 51.2 liters.

6. The dimensions on this welded square box are inside dimensions. Find the number of liters that the tank can hold. Round the answer to tenths. __________

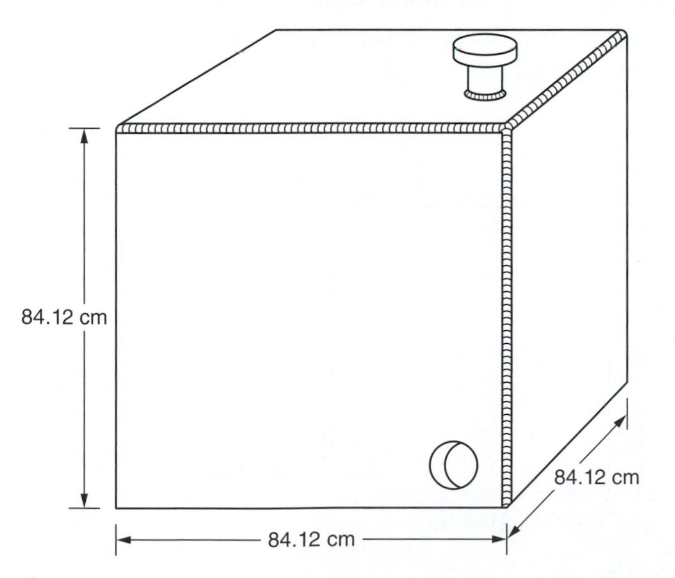

NOTE: Use the following information for Problems 7 and 8.

Welded tanks A and B are made from ⅛-inch steel plate. Outside measurements are given.

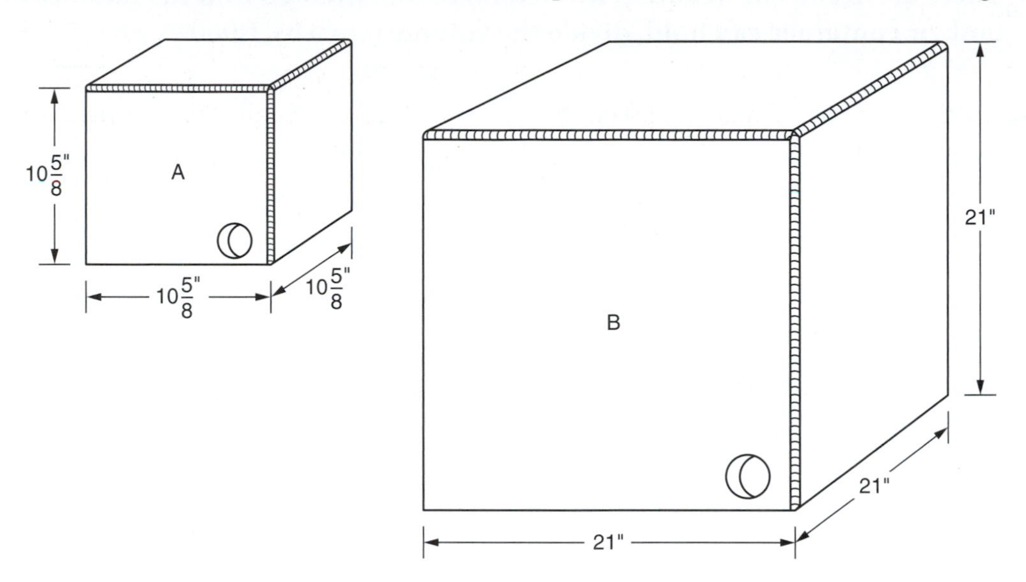

7. Find the volume of tank A. Round the answer to the nearest tenth cubic inch. __________

8. Find the volume of tank B. __________

9. The dimensions of welded storage tanks C and D are inside dimensions. The dimensions of tank D are exactly twice those of tank C. Is the volume of tank D twice the capacity of tank C? __________

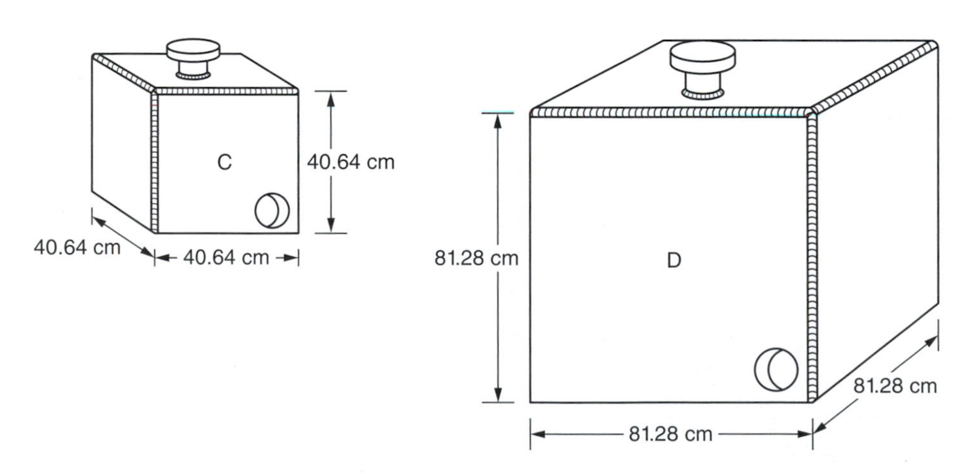

10. Cubed tanks A, B, C, and D are welded and filled with a liquid. Which of the tanks has a volume closest to one gallon? The dimensions are inside dimensions. ______

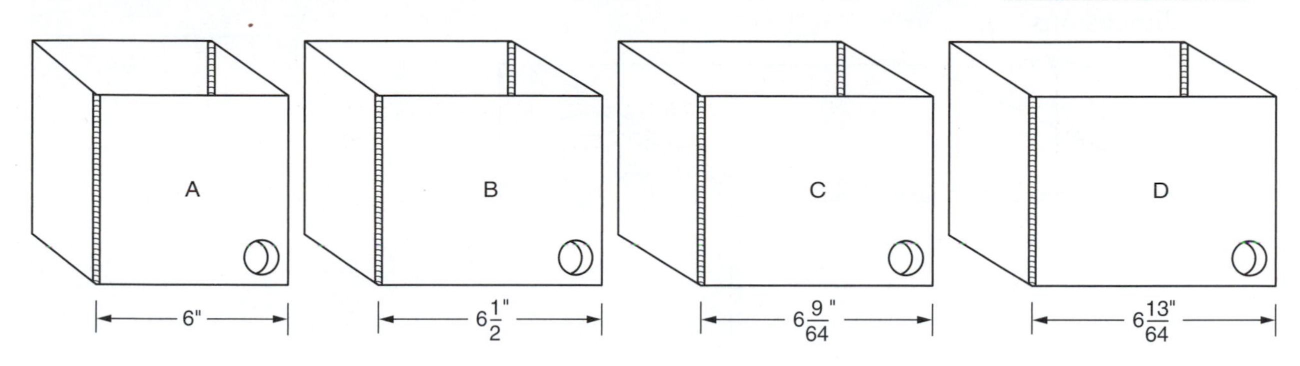

11. Nine fuel storage tanks for pickup trucks are welded. The dimensions are inside dimensions.

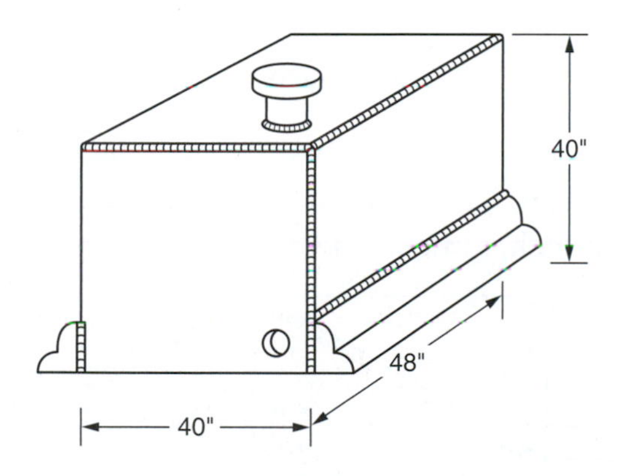

a. What is the total volume, in cubic inches, of the entire order of tanks? a. ______

b. What is the total volume in cubic feet? b. ______

12. A rectangular tank is welded from ⅛-inch steel plate to fit the specifications shown. How many gallons does the tank hold? Round the answer to three decimal places. The dimensions are inside dimensions. __________

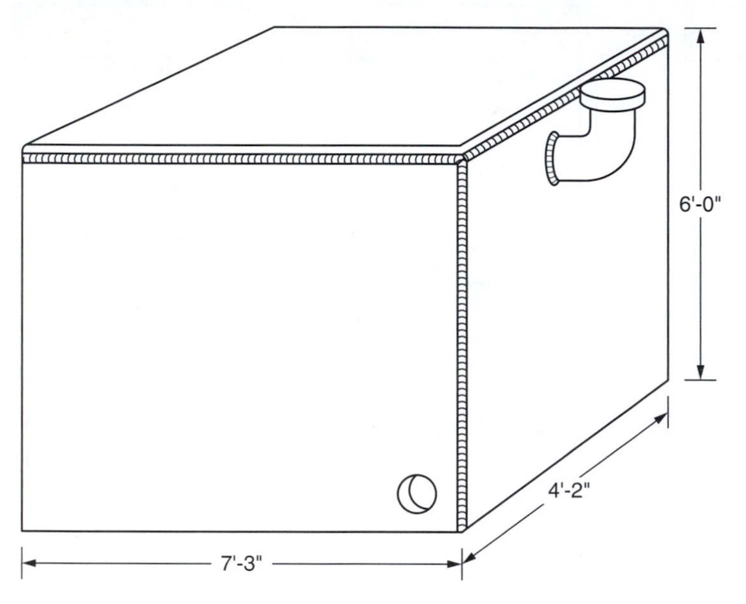

13. This welded tank has two inside dimensions given. The tank holds 80.53 gallons of liquid. Find, to the nearest tenth inch, dimension x. __________

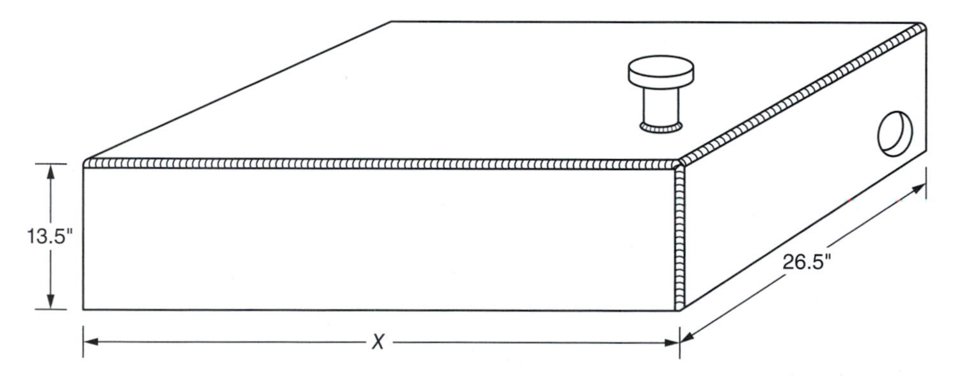

14. This rectangular welded tank is increased in length, so that the volume, in gallons, is doubled. What is the new length (dimension x) after the welding is completed? __________

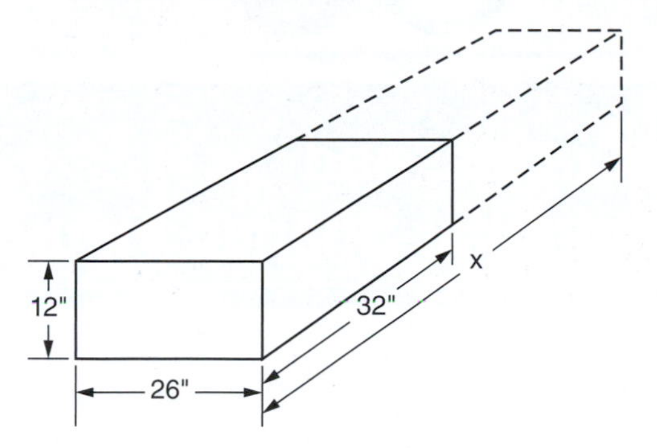

15. A pickup truck tank holds 89 liters of gasoline. Two auxiliary tanks are constructed to fit into spaces under the fenders of the truck. What is the total volume of the two tanks plus the original tank? __________

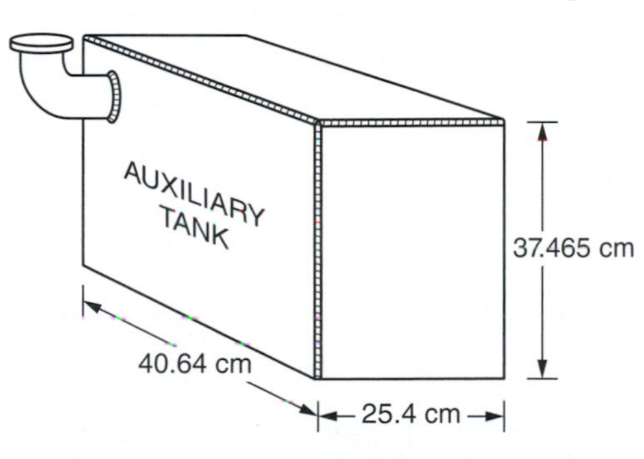

16. This welded steel tank is damaged. The section indicated is removed and a new bulkhead welded in its place. How many fewer liters will the tank hold after the repair? __________

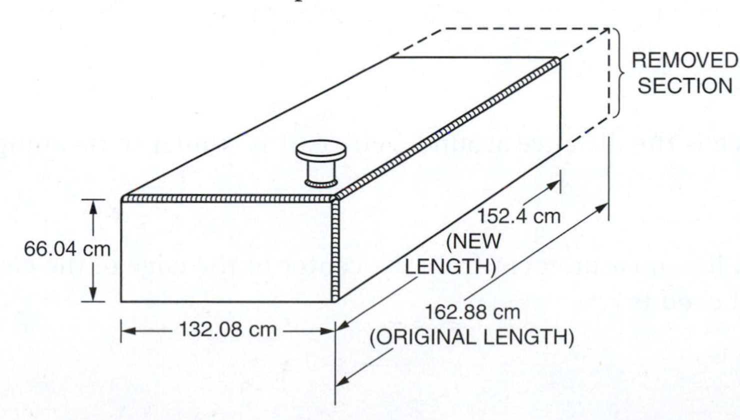

UNIT 30

Circumference of Circles, and Perimeter of Semicircular-Shaped Figures

Basic Principles

Definition of a circle and the parts of a circle.

Circle: A circle is a closed curved object, all parts of which are equally distant from the center.

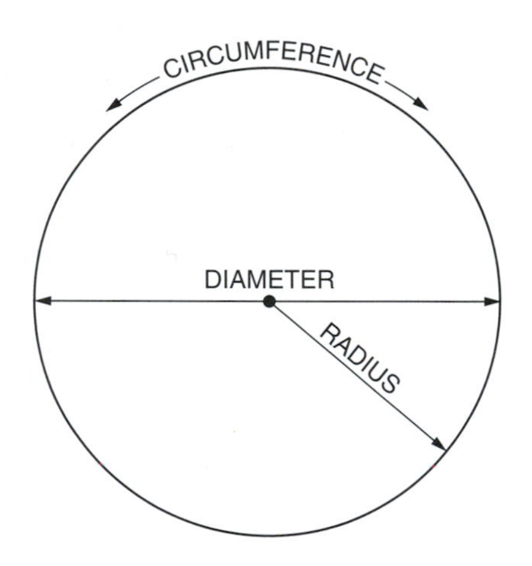

Circumference: Circumference is the distance around a circle: it is similar in meaning to perimeter. **Symbol used is *C*.**

Radius: The radius is a straight line measurement from the center to the edge of the circle; it is one-half the diameter. **Symbol used is *r*.**

Diameter: The diameter is a straight line through the center of the circle, traveling from edge to edge. It divides the circle in half, and is equal in length to 2 radii. **Symbol used is (*D*). Diameter is designated on blueprints with the symbol Ø.**

pi: The circumference of any circle is 3.1416 times the diameter of that circle. The number 3.1416 is represented by the Greek letter "pi". **The symbol used is π. Welding shops round π to 3.14.**

The formula for calculating the circumference of a circle is

$$C = \pi D$$

EXAMPLE: Using chalk and a rule, a circle with a diameter of 12″ is marked on steel plate. How many inches does the chalk travel in drawing a complete circle?

SOLUTION:

$$\begin{aligned} C &= \pi D \\ &= 3.14 \times 12'' \\ &= 37.68'' \end{aligned}$$

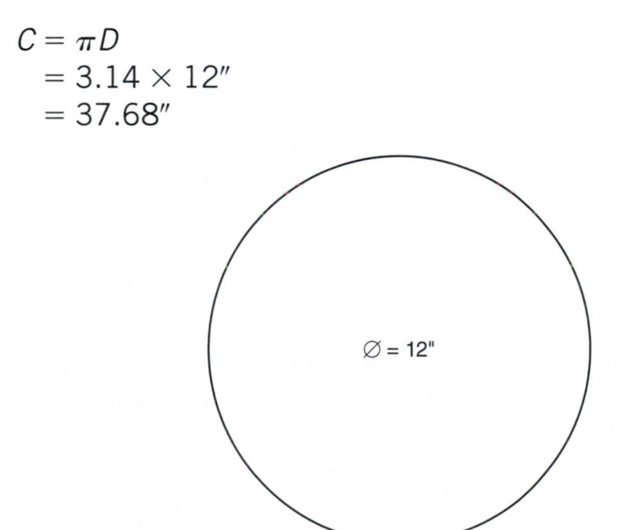

The chalk travels 37.68″.

Practical Problems

1. Circles A and B are cut from ⅜ steel plate. What is the circumference of both circles in inches? In feet?

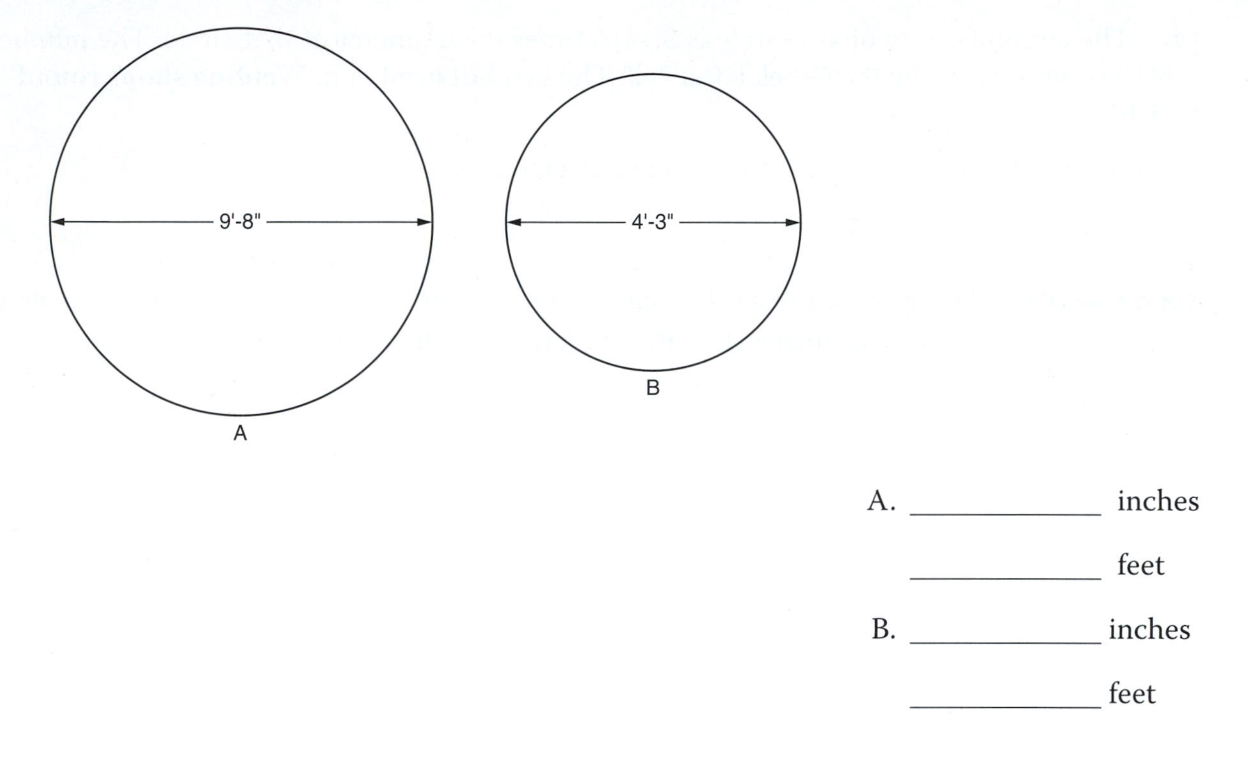

A. ____________ inches

____________ feet

B. ____________ inches

____________ feet

2. What is the circumference of a circle that has a radius of 6.3 cm? ____________

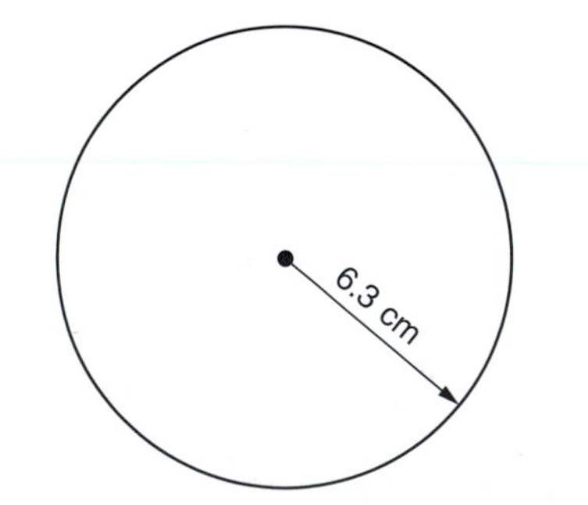

3. The Earth has an average diameter of approximately 7,914 miles. What is its average circumference? ____________

NOTE: Semicircular-shaped objects (half circles)

The measurement around (perimeter) a semicircular object is

a. $\frac{1}{2}$ the circumference of the circle, and

b. addition of the measurement of the diameter.

EXAMPLE: What is the distance around this semicircular figure?

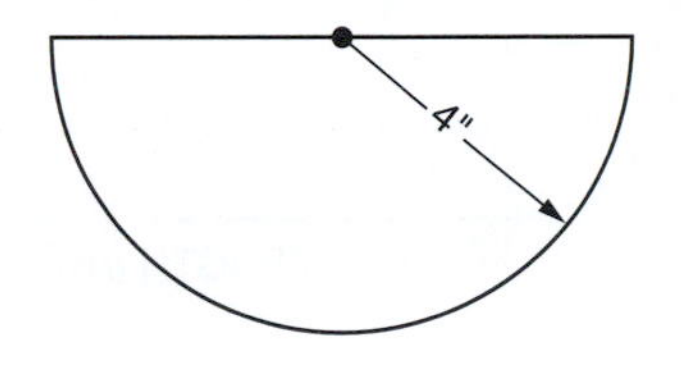

Procedure

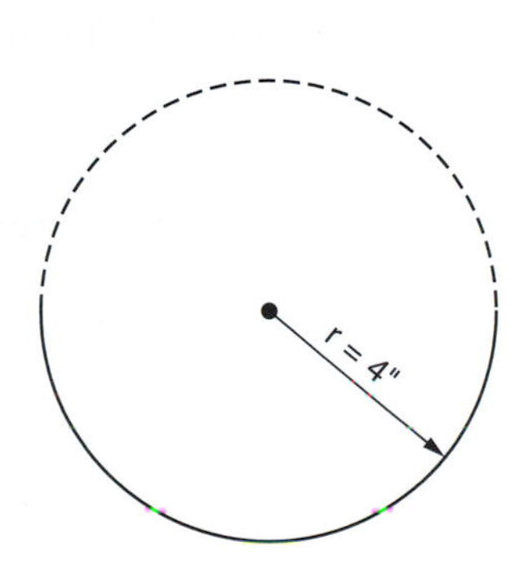

With a given radius of 4″, the diameter is 2 × 4″, or 8″.

$$C = \pi D$$
$$= 3.14 \times 8''$$
$$= 25.12''$$

$$\frac{1}{2} \text{ of } 25.12'' = 12.56''$$

SOLUTION:

```
  12.56
+  8.00   (measurement of the diameter)
  20.56″
```

ANSWER: The perimeter of the semicircular figure is 20.56″.

NOTE: Use this information for Problems 4 and 5. The perimeter of a semicircular-sided form is

a. the circumference of the circle formed by the 2 semicircular ends, plus

b. the addition of the measurement of 2 lengths (ℓ).

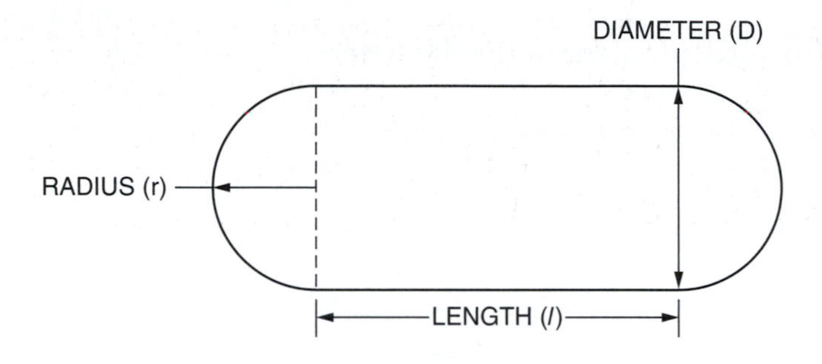

4. A semicircular-sided tank is welded in a shop. The bottom is cut from ⅛-inch plate with the dimensions shown. How long is the piece of metal used to form the sides of the tank? ____________

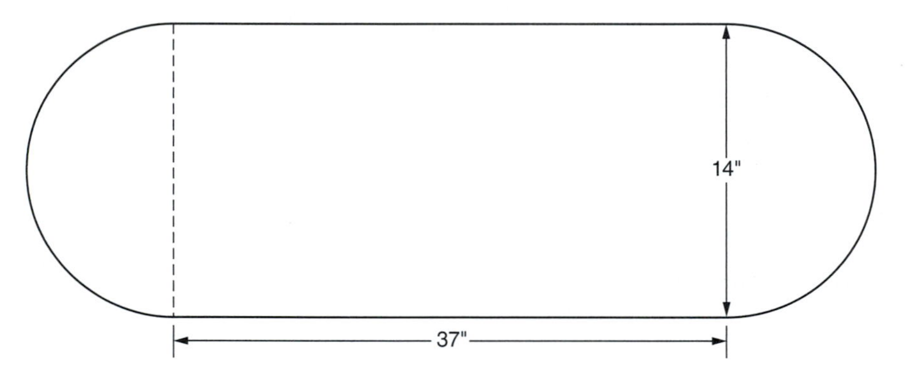

5. Find the distance around this semicircular-sided tank. Round the answer to three decimal places. ____________

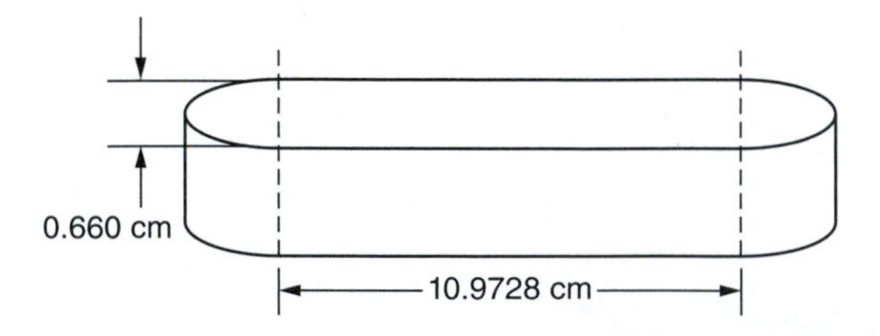

UNIT 31

Area of Circular and Semicircular Figures

Basic Principles

The formula for the area of a circle is as follows:

$$A = \pi r^2$$

Reminders

$\pi = 3.14$
$r^2 = r \times r$
$\varnothing$ = diameter

Determine the area of a circle with a radius of 8″.

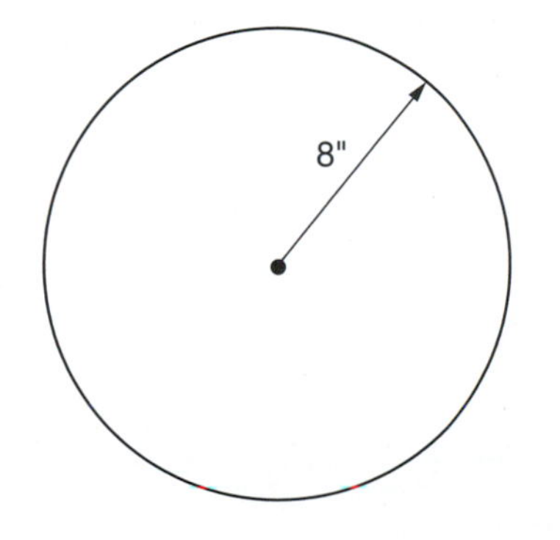

SOLUTION:
$$A = \pi r^2$$
$$= (3.14)(8 \times 8)$$
$$= 3.14 \times 64$$
$$= 200.96 \text{ in}^2$$

ANSWER: $A = 200.96 \text{ in}^2$

Practical Problems

1. A steel tank is welded as shown. Find, in square inches, the area of the circular steel bottom. ____________

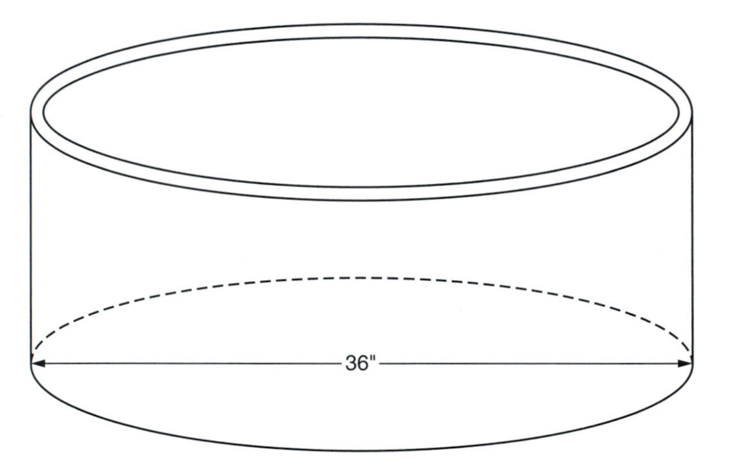

2. What is the area of a circle that has a diameter of 29 cm? Express the answer in cm^2, in^2, ft^2, m^2.

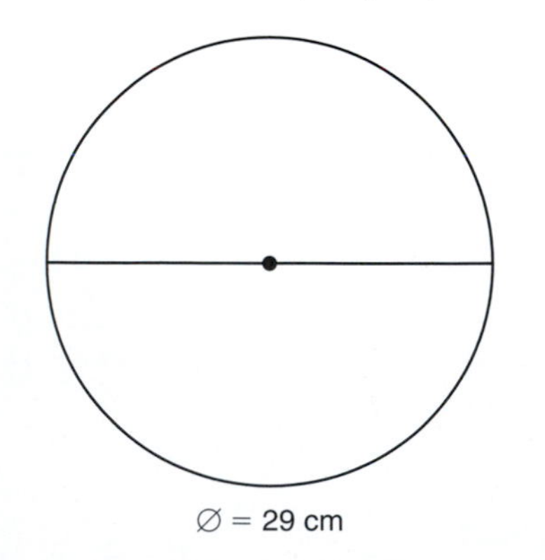

cm^2 ____________

in^2 ____________

ft^2 ____________

m^2 ____________

Area of Semicircular Figures

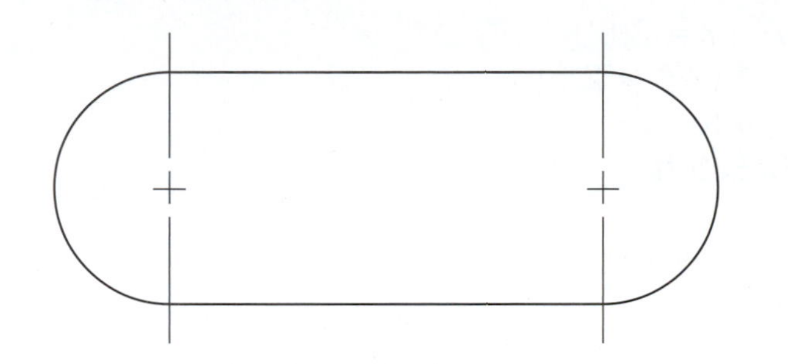

RULE: The area of this semicircular-sided piece of steel is equal to the sum of the areas of the two semicircles plus the rectangle.

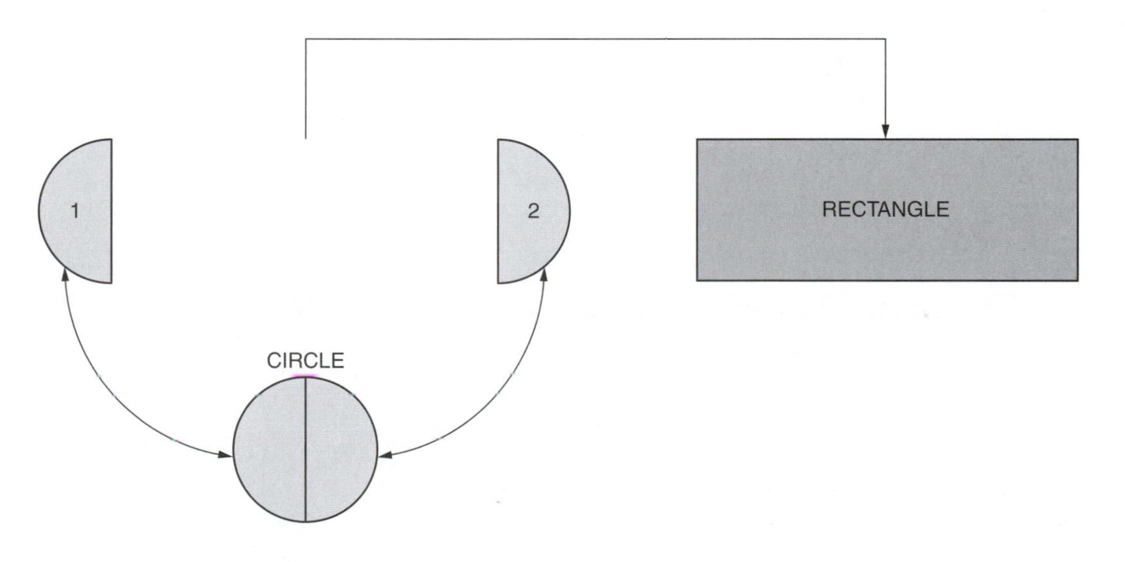

EXAMPLE: Determine the area of the following semicircular shape:

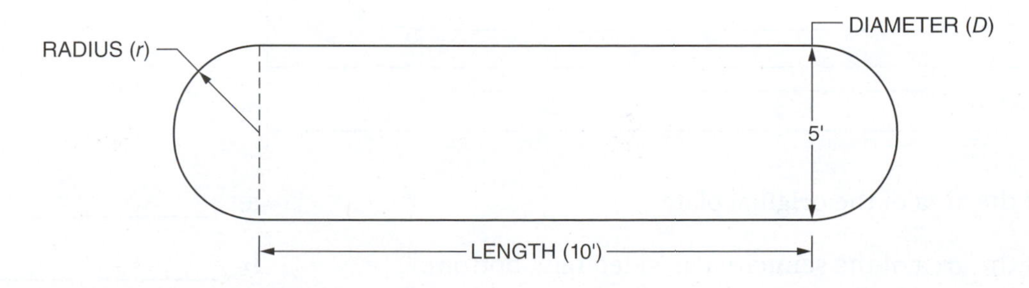

STEP 1 Area of circle.

$$A = \pi r^2 \quad r = 2.5'$$
$$= 3.14\ (2.5 \times 2.5)$$
$$= 3.14\ (6.25)$$
$$= 19.625\ \text{ft}^2$$

STEP 2 Area of rectangle.

$$A = \ell w$$
$$= 10' \times 5'$$
$$= 50\ \text{ft}^2$$

STEP 3

$$\begin{array}{r} 19.625\ \text{ft}^2 \\ +50.000\ \text{ft}^2 \\ \hline 69.625\ \text{ft}^2 \end{array}$$

ANSWER: $A = 69.625\ \text{ft}^2$

Practical Problems

3. This tank bottom is cut from 3⁄16″ steel plate. Express each answer in square inches.

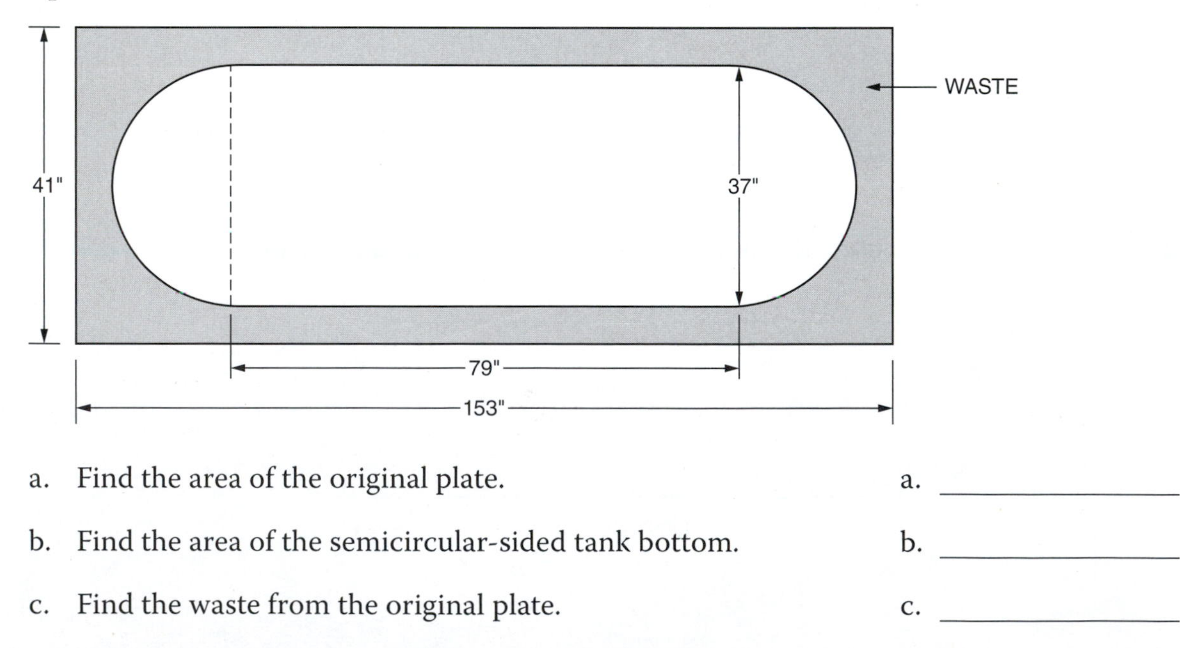

a. Find the area of the original plate. a. ______________

b. Find the area of the semicircular-sided tank bottom. b. ______________

c. Find the waste from the original plate. c. ______________

NOTE: Use this information for Problems 4–6.

Find the area of each semicircular-sided tank bottom. Express each area in square inches.

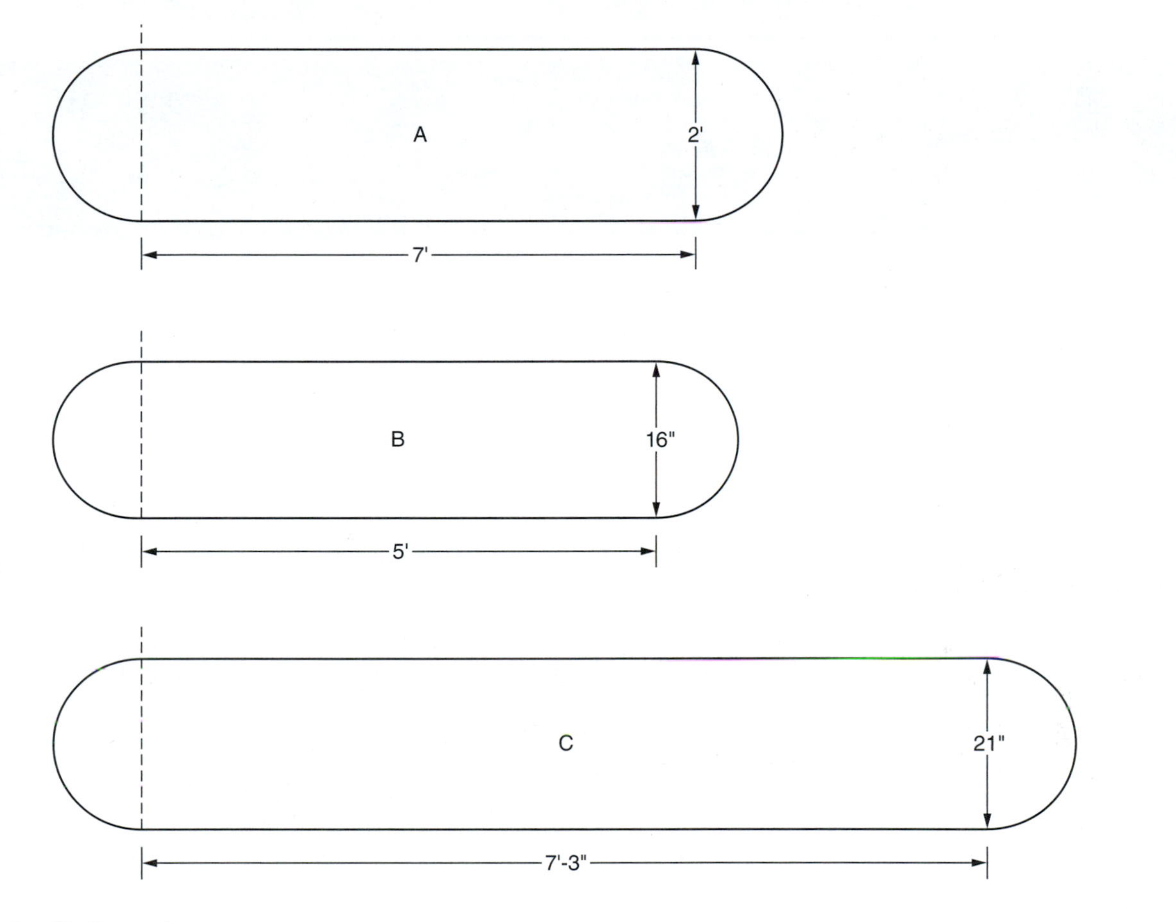

4. Bottom A __________

5. Bottom B __________

6. Bottom C __________

UNIT 32

Volume of Cylindrical Shapes

Basic Principles

Formula: The formula for the volume of a cylinder is as follows:

$$V = (\pi r^2)h$$

NOTE: (πr^2) is the formula for the area of the cylinder face. When this is multiplied by the height of the cylinder, the volume is found.

EXAMPLE: What is the volume of a cylinder with a radius of 6″ and a height of 9″?

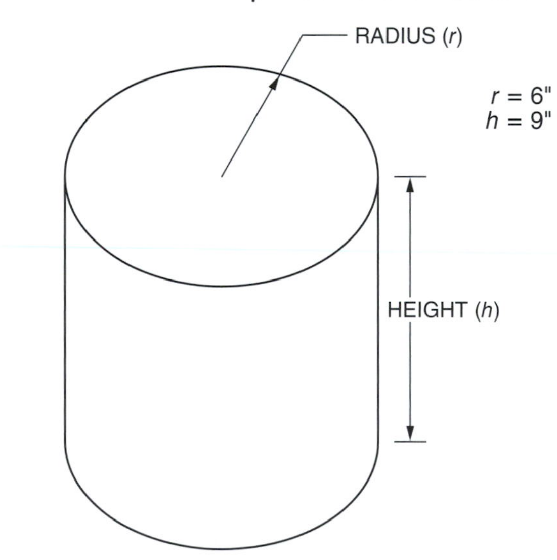

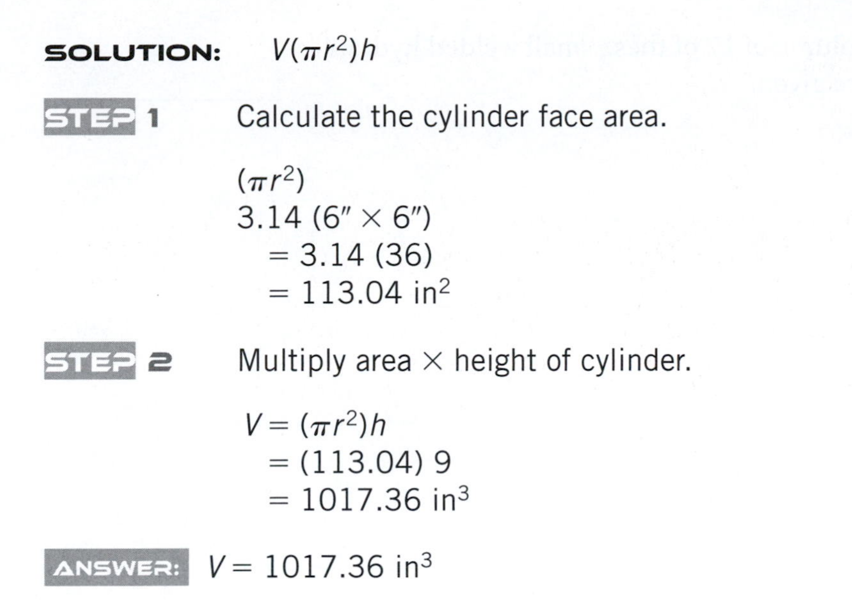

SOLUTION: $V(\pi r^2)h$

STEP 1 Calculate the cylinder face area.

(πr^2)
$3.14\ (6'' \times 6'')$
$= 3.14\ (36)$
$= 113.04\ \text{in}^2$

STEP 2 Multiply area × height of cylinder.

$V = (\pi r^2)h$
$= (113.04)\ 9$
$= 1017.36\ \text{in}^3$

ANSWER: $V = 1017.36\ \text{in}^3$

Practical Problems

Find, in cubic inches, the volume of each piece of round stock.

1. $D = 10$ in; $h = 60$ in __________
2. $D = 48$ in; $h = 48$ in __________
3. $D = 8.125$ in; $h = 59.875$ in __________
4. $D = 10.625$ in; $h = 72.75$ in __________

Find, in cubic feet, the volume of each cylinder.

5. $r = 12$ in; $h = 48$ in __________
6. $r = 3$ in; $h = 120$ in __________
7. $D = 12$ in; $h = 24$ in __________
8. $D = 8.375$ in; $h = 22.125$ in __________

9. Find, in cubic inches, the volume of 17 of these small welded hydraulic tanks. Inside dimensions are given. ______________

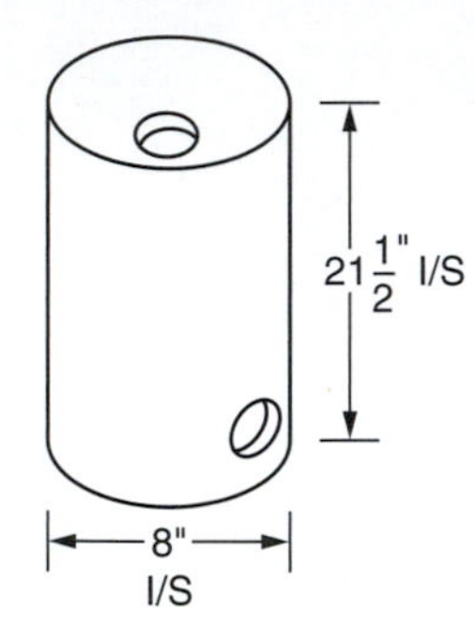

NOTE: Inside dimensions abbreviation: I/S.

Outside dimensions abbreviation: O/S.

Semicircular-Shaped Tanks and Solids

RULE: The volume of this semicircular-shaped solid is equal to the sum of the two semicylinders at the ends, and the rectangularly-shaped piece in the center.

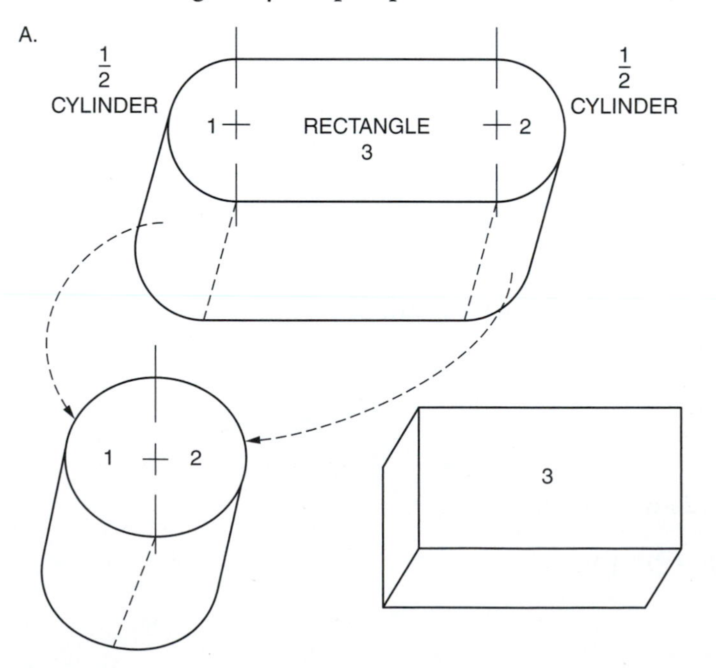

EXAMPLE: Find the volume of water that can be held in the semicircular-shaped tank. Dimensions given are inside dimensions.

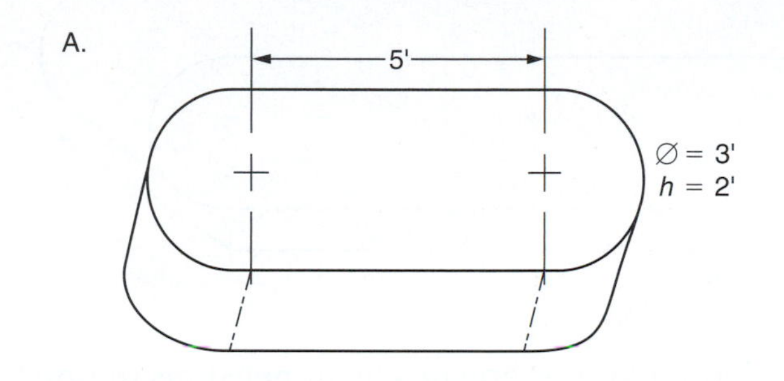

SOLUTION:

STEP 1 Volume of cylindrical ends.

$$
\begin{aligned}
V &= (\pi r^2)h \\
&= 3.14(1.5' \times 1.5')2 \\
&= 3.14(2.25)2 \\
&= 7.065 \times 2' \\
V &= 14.13 \text{ ft}^3
\end{aligned}
$$

STEP 2 Volume of rectangularly shaped center.

$$
\begin{aligned}
V &= \ell wh \\
&= 5' \times 3' \times 2' \\
&= 15 \times 2 \\
V &= 30 \text{ ft}^3
\end{aligned}
$$

STEP 3

$$
\begin{array}{r}
14.13 \text{ ft}^3 \\
+30.00 \text{ ft}^3 \\
\hline
44.13 \text{ ft}^3
\end{array}
$$

ANSWER: 44.13 cubic feet of water can be held in tank A.

Reminder: All dimensions must be in the same unit of measure before calculating: inches and inches, feet and feet, and so on.

10. What is the volume of this semicircular-sided solid in cubic feet? ______________

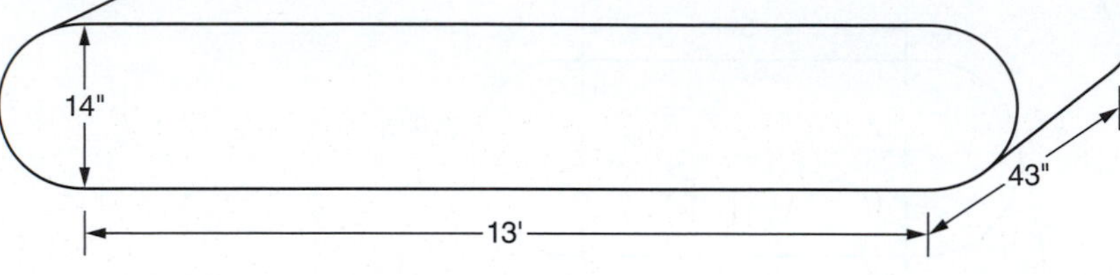

11. Two semicircular-sided tanks are shown. The dimensions of one tank are exactly twice the dimensions of the other tank. Is the volume of the larger tank twice the volume of the smaller tank? Explain volume comparison ______________

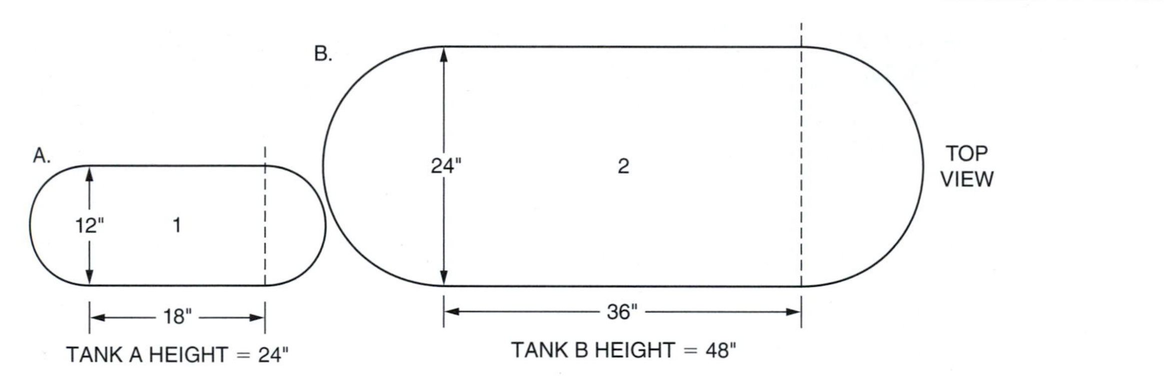

UNIT 33

Volume of Cylindrical and Complex Containers

Basic Principles

The volume of cylindrical containers is found using inside dimensions of the pipe or containers. If outside dimensions are given, subtract wall thicknesses as a first step. Review volume formulas from previous chapters.

Practical Problems

1. A pipe with an outside diameter of 10 inches is cut into 3 pieces. Find the volume of each piece, in cubic inches. Pipe wall thickness is .5″.

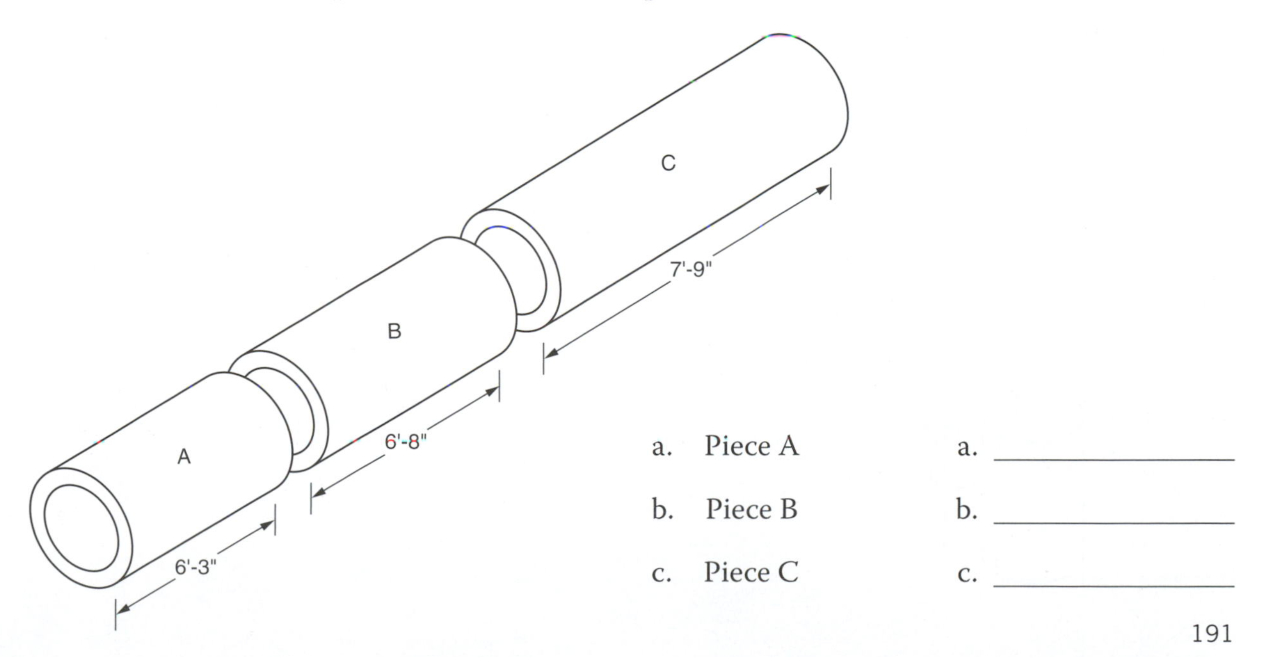

a. Piece A　　　a. ____________

b. Piece B　　　b. ____________

c. Piece C　　　c. ____________

2. An outside storage tank is welded. The dimensions given are inside dimensions.

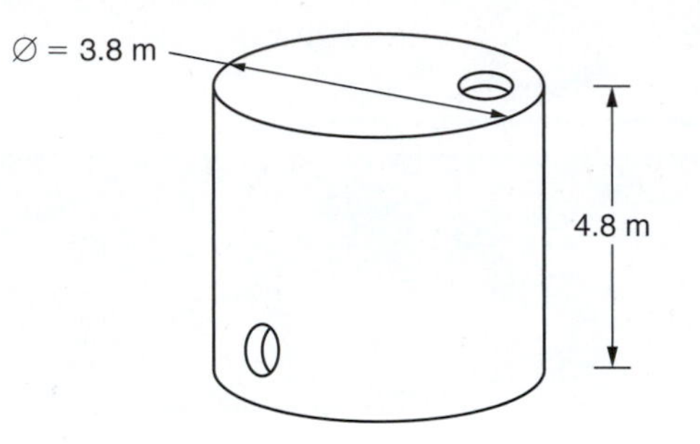

a. Find, in cubic meters, the volume of the tank. a. ________________

b. Find, in liters, the volume of the tank. b. ________________

3. The dimensions on these three cylindrical welded tanks and connecting pipes are inside dimensions. The tanks are connected as shown and are filled with liquid. The system is completely filled, including the connecting pipes. What is the total volume of the system? ________________

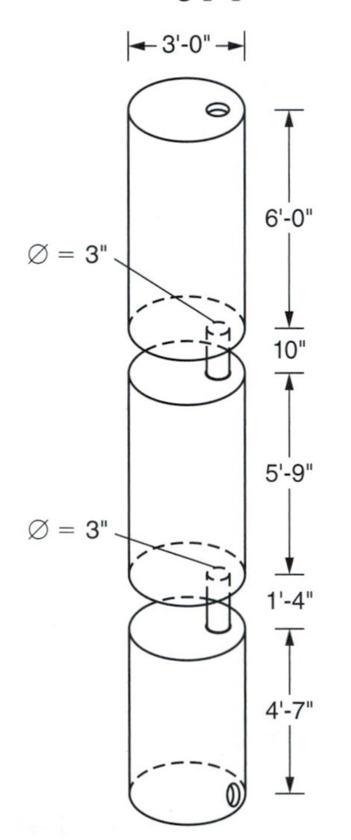

4. A 90° two-piece elbow is cut and welded from a 60.96-cm inside diameter pipe. Find the volume of the elbow to the nearest hundredth cubic meter. ____________

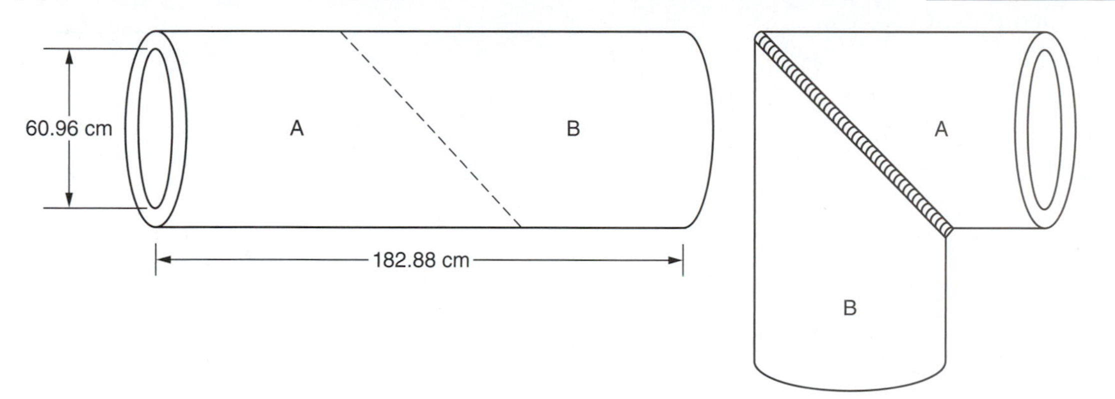

5. Two settling tanks are welded together. The dimensions given are inside dimensions. Find, in gallons, the volume of the entire system. Round the answer to the nearest tenth gallon. ____________

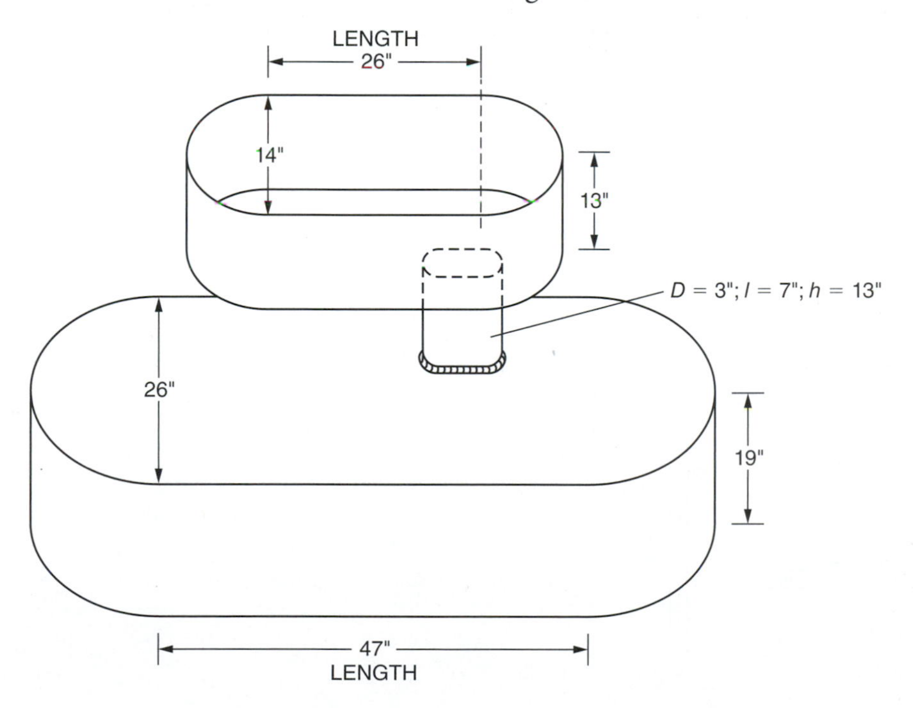

6. A length of welded irrigation pipe has the dimensions shown. Twenty of these lengths are welded together. What is the total volume of the welded pipes? ____________

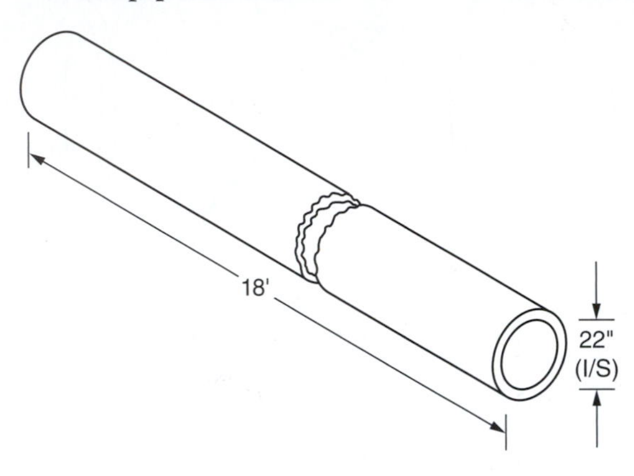

7. A weldment consisting of a semicircular-sided tank and a steel angle frame is constructed as shown. The dimensions given are inside dimensions. What is the volume of the complete tank to the nearest gallon? ____________

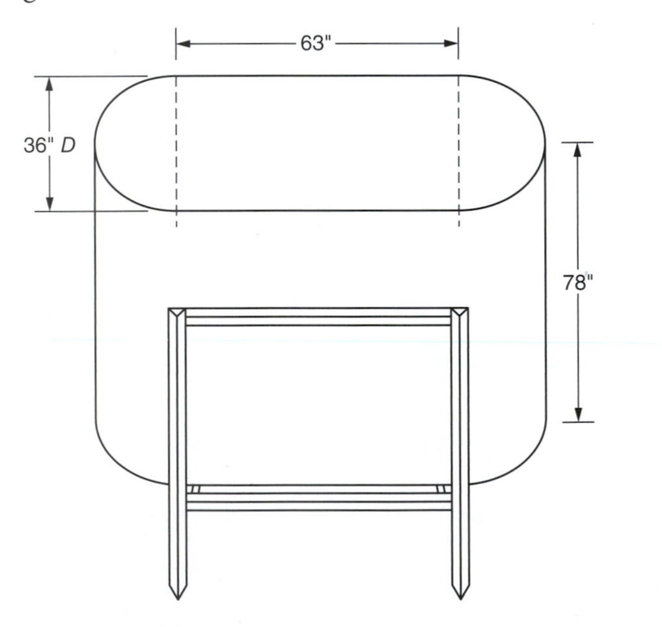

8. Using semicircular-sided pipes, this manifold system is welded. The dimensions given are inside dimensions.

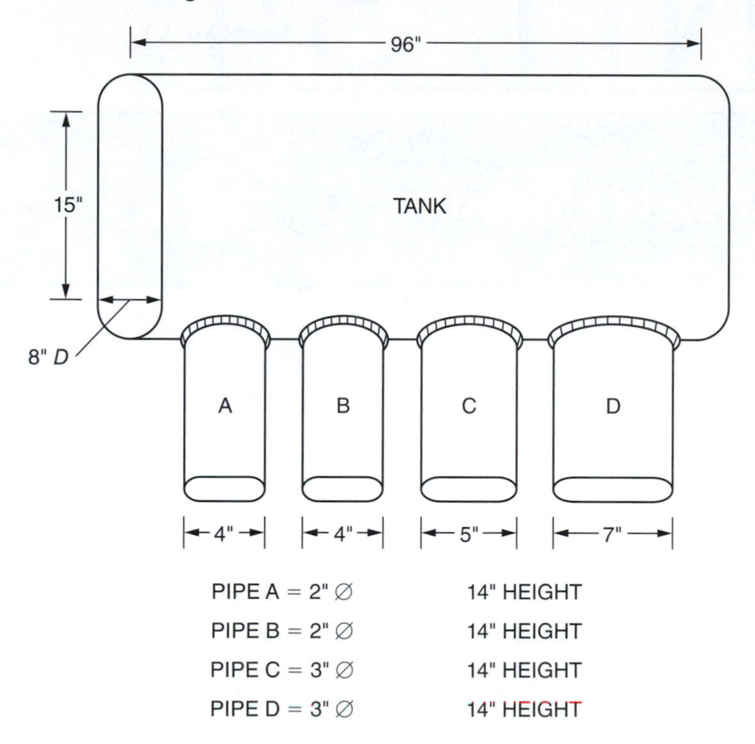

a. Find, in cubic inches, the total volume of the pipes. a. ____________

b. Find, in cubic inches, the volume of the tank. b. ____________

c. Find, in gallons, the volume of the entire manifold system. c. ____________

UNIT 34

Mass (Weight) Measure

Basic Principles

English Measure Note

1 in^3 of steel = .2835 lb. (3.5273 in^3 steel = 1 pound)

Metric Measure Note

1 cm^3 of steel = 7.849 grams (may vary per resource)

RULE: To determine the weight of a steel piece, calculate the quantity of in^3 or cm^3, and multiply times the appropriate figure.

EXAMPLE 1: A 9″ piece of steel round stock has a diameter of 4″. Calculate the weight.

SOLUTION: Volume × weight

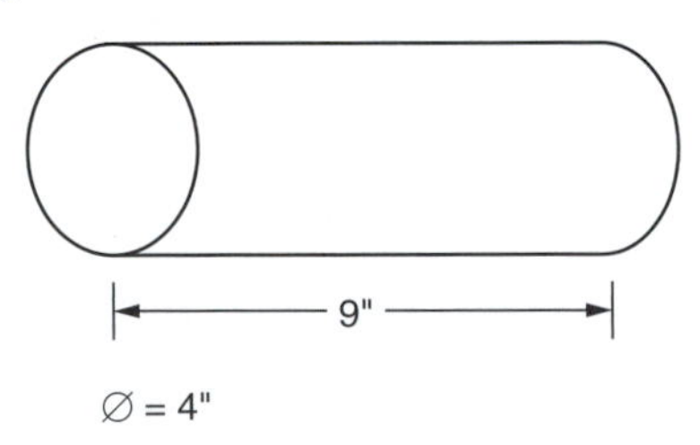

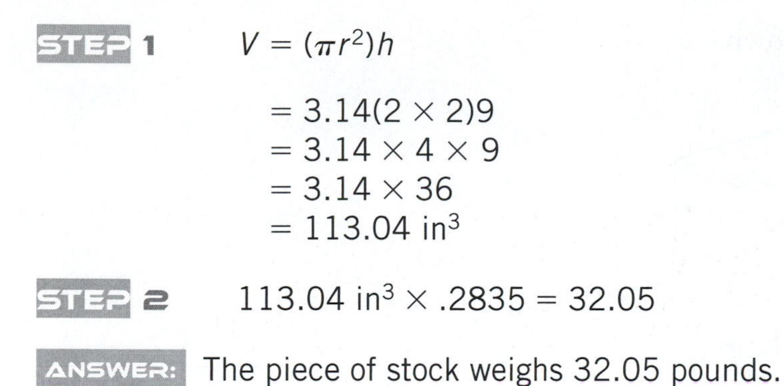

STEP 1 $V = (\pi r^2)h$

$= 3.14(2 \times 2)9$
$= 3.14 \times 4 \times 9$
$= 3.14 \times 36$
$= 113.04 \text{ in}^3$

STEP 2 $113.04 \text{ in}^3 \times .2835 = 32.05$

ANSWER: The piece of stock weighs 32.05 pounds.

Practical Problems

1. Fourteen pieces of cold-rolled steel shafting are cut as shown. What is the total weight of the 14 pieces of steel in pounds? ____________

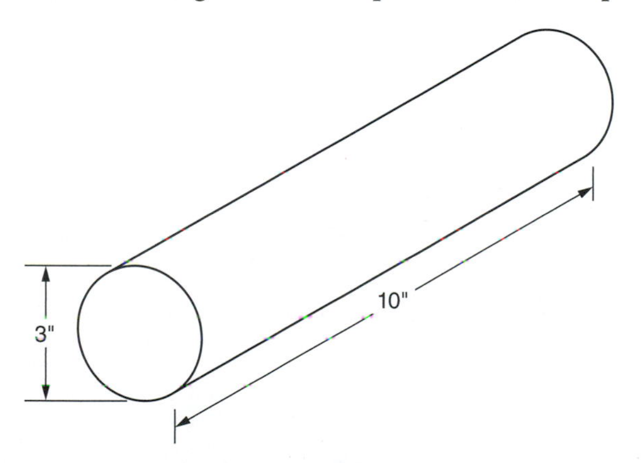

2. Steel angle legs for a tank stand have the dimensions shown. Find the weight of 20 legs in kilograms. ____________

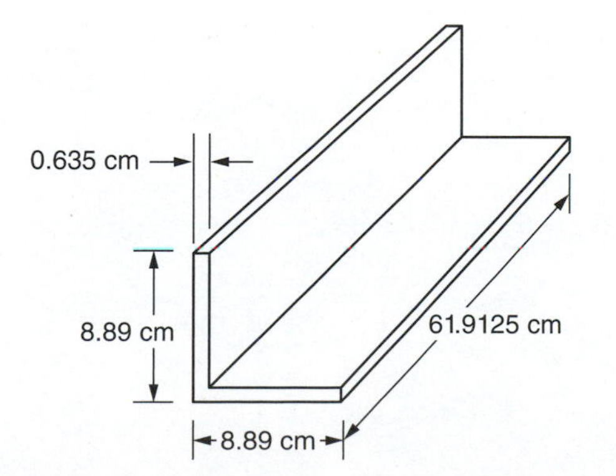

3. A circular tank bottom is cut as shown.

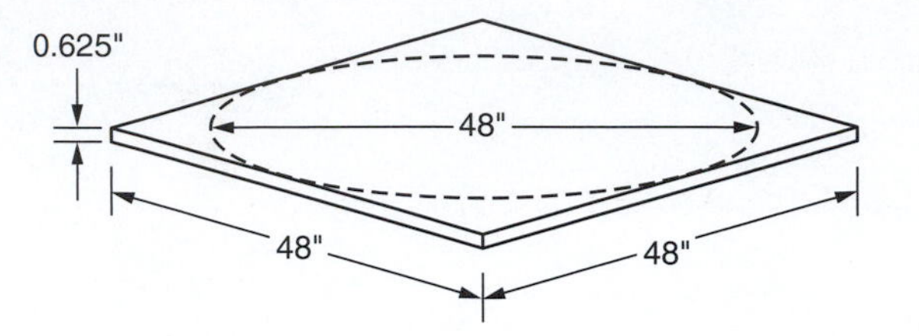

a. Find the weight of the circular bottom. a. ________________

b. Find the weight of the wasted material. b. ________________

4. An open-top welded bin is made from ¼-inch plate steel. What is the total weight, in pounds, of the 5 pieces used for the bin? ________________

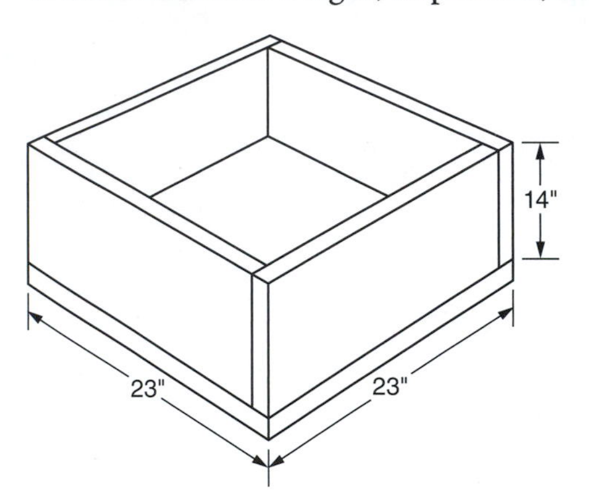

5. Sixteen circular blanks for sprockets are cut from 1.27-cm plate.

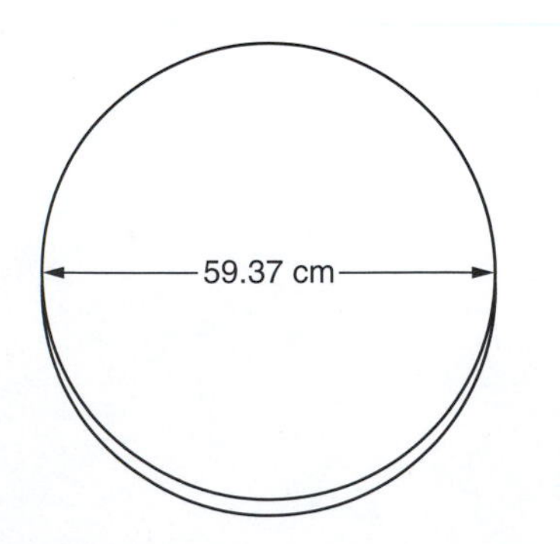

a. What is the weight of 1 blank in kilograms? a. ______________

b. What is the weight of all of the blanks in kilograms? b. ______________

6. Pieces of ⅜-inch bar stock are used for welding tests. Find, in pounds, the weight of 1 piece of the bar stock. ______________

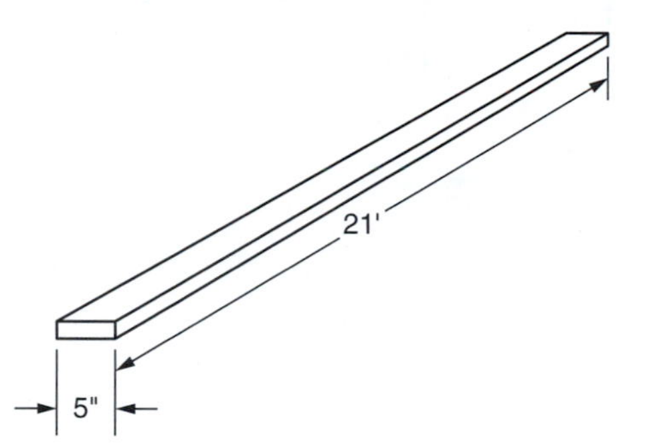

7. A column support gusset is shown. Find, in pounds, the weight of 52 of these gussets. ______________

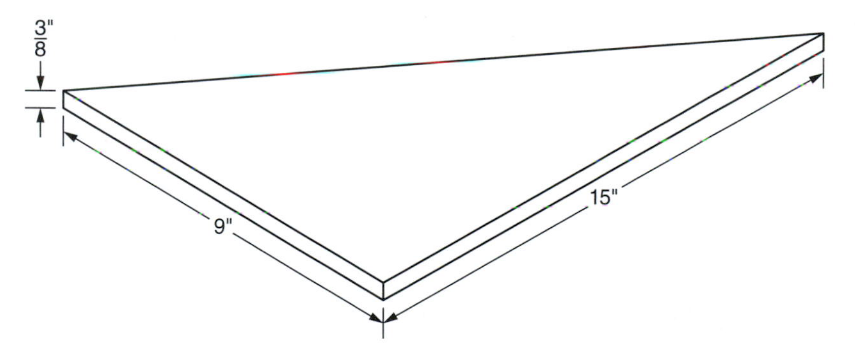

8. Find, in pounds, the weight of this adjustment bracket. ______________

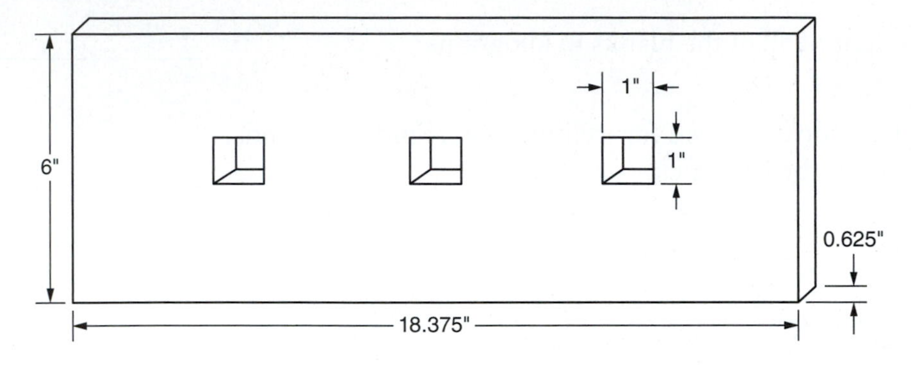

9. A welder flame-cuts 10 roof columns from pipe as shown. Find the total weight of the columns (O/S = outside diameter). ______________

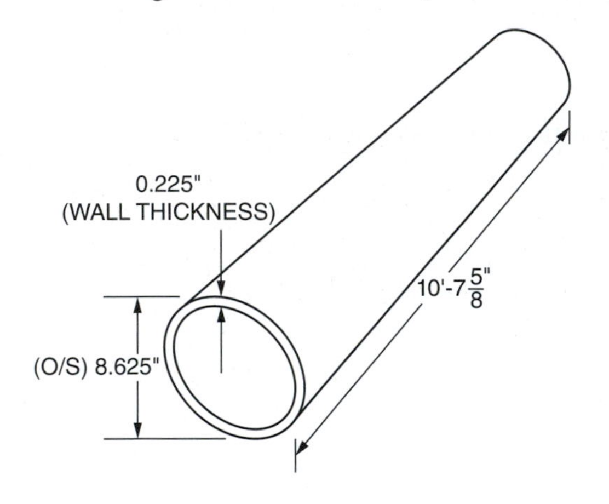

SECTION 7

ANGULAR DEVELOPMENT AND MEASUREMENT

UNIT 35

Angle Development

Basic Principles

Angles are formed and measured at the center of a circle. The radius, fixed in the center, rotates inside the circle. This movement is similar to the second hand on a clock.

In one full revolution, the radius moves 360 small increments. Each increment is called a "degree."

Figures A–F demonstrate angle development as the radius moves one revolution in a circle.

RULE: There are 360 degrees in a circle and 180 degrees on each side of a straight line. (See the diameter in Figure D.)

A 90-degree angle is called a "right angle." (See Figure C.)

Symbols

Symbol for degree: (°).

Symbol for angle: (∠).

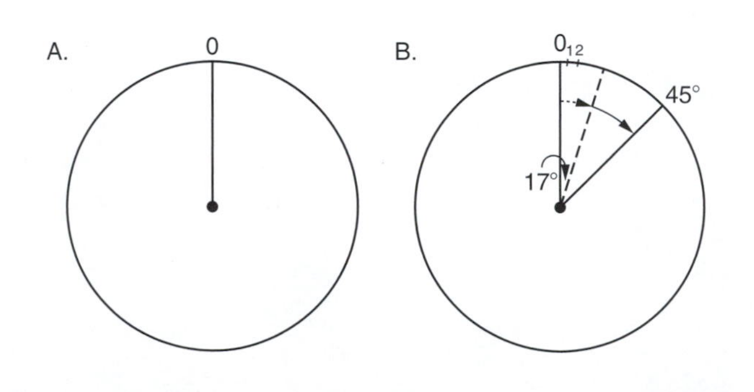

When the radius moves ¼ revolution, a 90° angle is formed. A box symbol (⌐) placed in the angle indicates a 90° angle. Placement of degrees inside the circle show the actual angle measurement. Degrees written on the outside of the circle, as in Figures A, B, D, E, and F, describe angles and are not linear measurements.

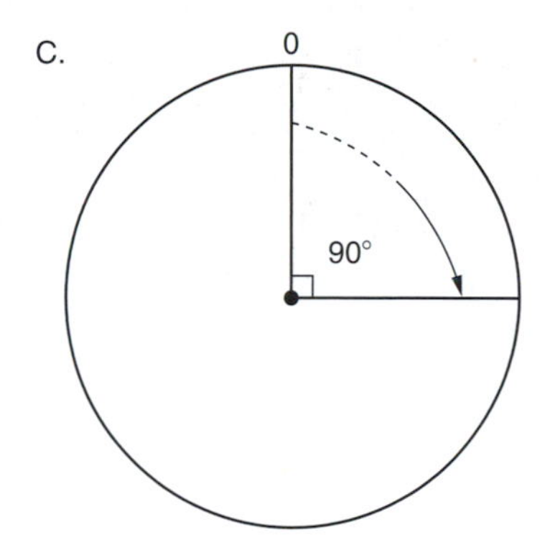

As the radius revolves, larger angles are formed:

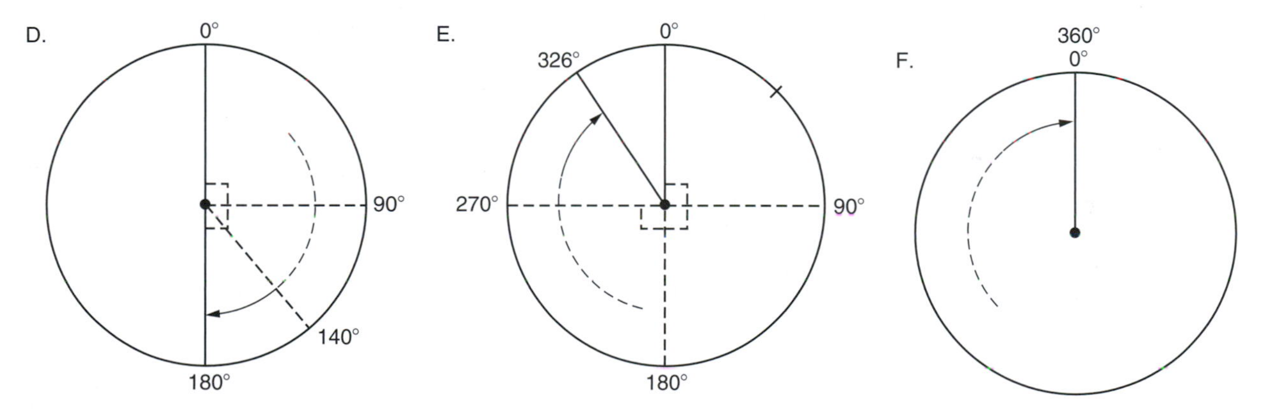

Each degree is divided into 60 small units called "minutes," and each minute is divided into 60 smaller units called "seconds."

Symbol for minutes: (') is similar to the symbol for feet.

Symbol for seconds: (") is similar to the symbol for inches.

EXAMPLE: $\angle A = 63° 15' 38"$
(Angle A is 63 degrees, 15 minutes, and 38 seconds.)

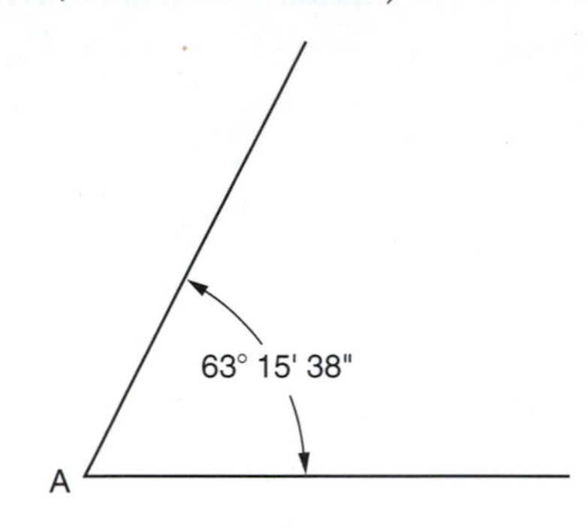

Angle A is larger than 63°, smaller than 64°.

NOTE: Each corner of a geometric shape, i.e. square, triangle, rectangle, etc., is formed from the center of a circle.

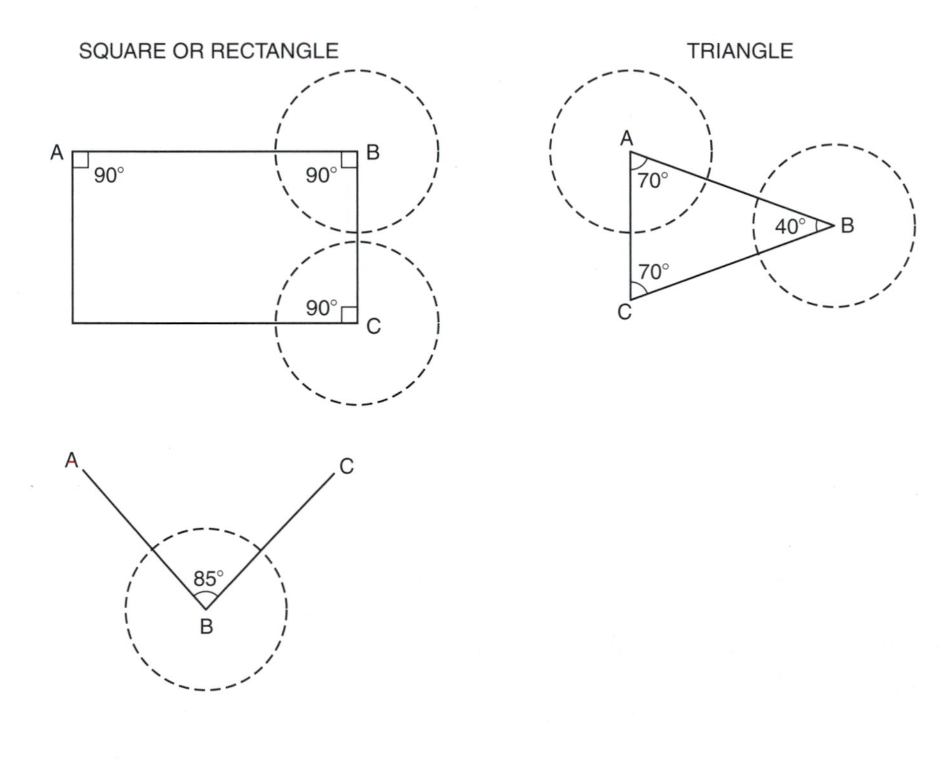

PROBLEM: How many degrees are in ½ circle?

Reminder

A full circle equals 360°.

SOLUTION: ½ of 360°

$$\frac{1}{2} \times 360 \rightarrow \frac{1}{2} \times \frac{360}{1} \rightarrow \frac{1}{\cancel{2}_{1}} \times \frac{\cancel{360}^{180}}{1} \rightarrow \frac{180}{1}$$

ANSWER: $\frac{1}{2}$ circle $= 180°$

Practical Problems

How many degrees are in each of these parts of a circle?

1. $\frac{1}{3}$ circle ________
2. $\frac{3}{4}$ circle ________
3. $\frac{5}{6}$ circle ________
4. $\frac{1}{16}$ circle ________

PROBLEM: 180° is what part of a circle?

SOLUTION: Set up as a fraction, and reduce if possible.

$$\frac{180}{360} \rightarrow \frac{180 \div 90}{360 \div 90} \rightarrow \frac{\cancel{180}^{2}}{\cancel{360}_{4}} \rightarrow \frac{\cancel{180}^{\cancel{2}^{1}}}{\cancel{360}_{\cancel{4}_{2}}} \rightarrow \frac{1}{2}$$

ANSWER: 180° is $\frac{1}{2}$ of a circle.

What part of a circle are the following angular measurements?

5. 60° ______________

6. 45° ______________

7. 90° ______________

8. 120° ______________

9. 160° ______________

10. This bolt hole circle is on a radius of 6″ and has four equally spaced holes. How many degrees apart are the hole centers? ______________

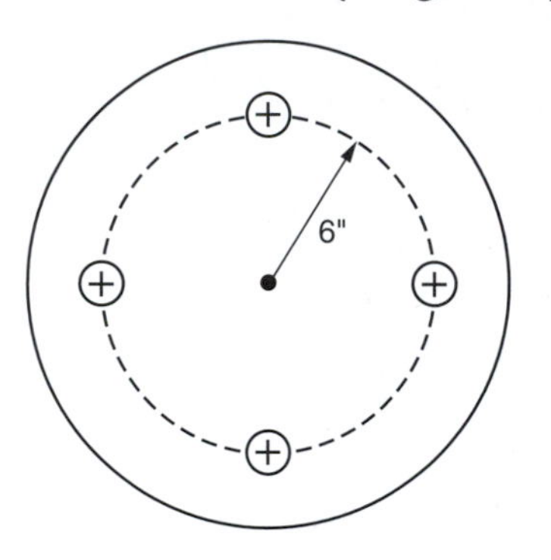

11. How many degrees apart are the centers of the holes on the bolt hole circle? ______________

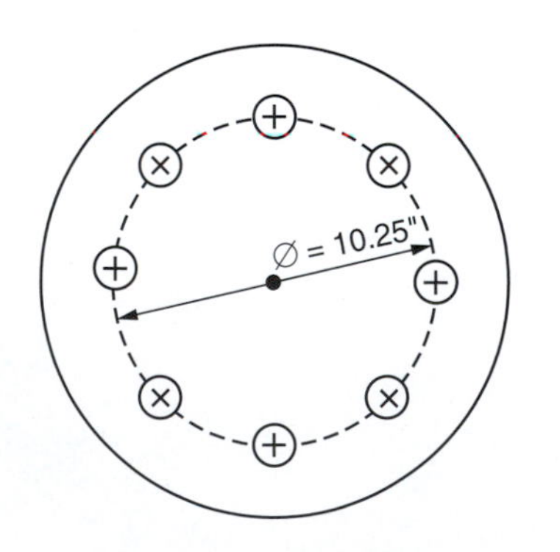

12. This pipe flange is drilled on a 4½-inch radius and on a 5-inch radius. How many degrees farther apart are the holes in the 10-inch circle than the holes in the 9-inch circle? __________

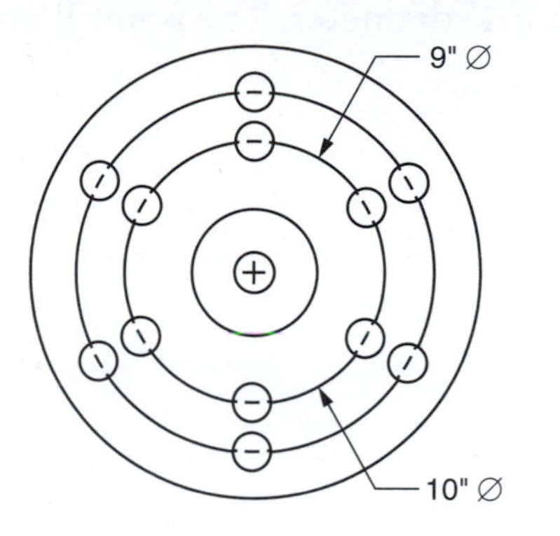

13. How many degrees apart are the equally spaced holes in each of these pipe flanges?

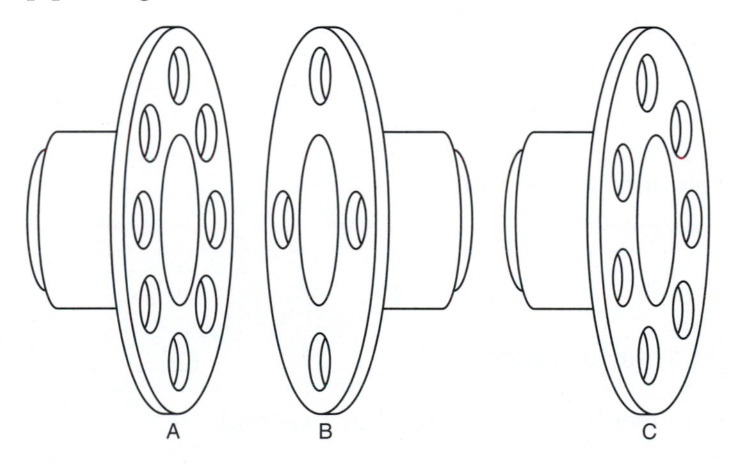

a. Flange A a. __________

b. Flange B b. __________

c. Flange C c. __________

Additional Practical Problems

Points A and B are marked on a propeller blade. The blade is moving at one revolution per second (RPS). The circle that point A makes in a revolution has a 4′ diameter. The point B circle has an 8′ diameter.

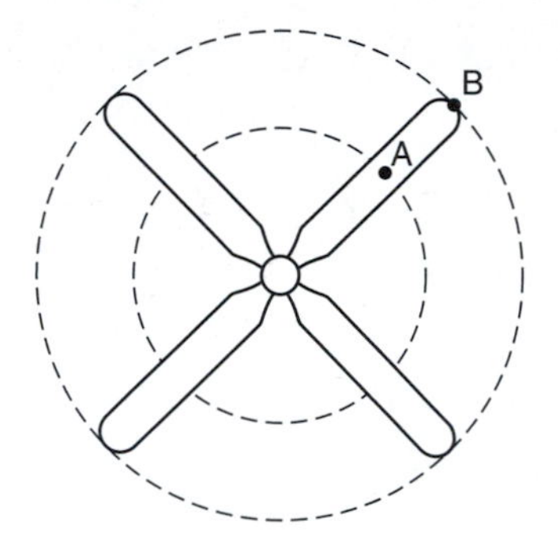

14. Which point is moving faster?

    a. A

    b. B

    c. the speed is the same for either ____________

15. Which point turns the most revolutions in one minute?

    a. A

    b. B

    c. the number of rotations is the same for either ____________

16. Determine the distance traveled in feet by points A and B in 1 second and in 1 hour.

    A ________ B ________

    A ________ B ________

17. Determine the speed of points A and B in feet per second and miles per hour.

    A ________ B ________

    A ________ B ________

UNIT 36

Angular Measurement

Basic Principles

Angles can be added, subtracted, multiplied, and divided. The following are examples of denominant numbers as explained in the Appendix, Section 1.

EXAMPLE 1: **Addition**

$$\begin{array}{rrr} 14^\circ & 32' & 15'' \\ +150^\circ & 13' & 22'' \\ \hline \end{array} \rightarrow \begin{array}{r|r|r} 14^\circ & 32' & 15'' \\ +150^\circ & 13' & 22'' \\ \hline 164^\circ & 45' & 37'' \end{array}$$

ANSWER: 164° 45' 37"

NOTE: All units are kept in separate columns. Dashed lines are used in the examples to help show separate degrees, minutes, and seconds columns.

EXAMPLE 2: **Addition**

$$\begin{array}{rrr} 36^\circ & 29' & 13'' \\ +114^\circ & 38' & 52'' \\ \hline \end{array} \rightarrow \begin{array}{r|r|r} 36^\circ & 29' & 13'' \\ +114^\circ & 38' & 52'' \\ \hline 150^\circ & 67' & 65'' \end{array}$$

NOTE: In our answer, we see that 65" contains 1 whole minute (1' = 60"). All minutes need to be moved from the seconds column into the minutes column.

RULE: Whenever seconds, minutes, or degrees move to the neighboring column, they change either from 60 to 1, or 1 to 60.

STEP 1 In Example 2, the 60 seconds are moved and added to the "minutes" column as 1 minute. 5" remain in the seconds column.

150° 67' 65" →

150°	67'	65"
	+1	−60"
	68'	5"

 150° 68' 5"

NOTE: 68 minutes contain 1 whole degree (1° = 60'). All minutes equaling whole degrees need to be moved and added to the degrees column.

150°	68'	5"
+ 1	−60'	
151°	8'	5"

ANSWER: 151° 8' 5"

Practical Problems

1. 93° 14' 10"
   +18° 59' 58"

2. 45° 30' 6"
   +19° 14' 0"

3. 180° 17' 0"
   +90° 45' 19"

EXAMPLE 3: Subtraction

90°	43'	12"
−25°	14'	6"

→

81	31	
~~9~~0°	~~4~~3'	12"
−25°	14'	6"
65°	29'	6"

ANSWER: 65° 29' 6"

EXAMPLE 4: Subtraction with Borrowing

180°	19'	14"
−90°	21'	3"

→

180°	79'	14"
−90°	21'	3"
		11"

NOTE: 21 cannot be subtracted from 19. Borrowing 1 degree will add 60 minutes. (1° = 60")

179°	1° 60'	
~~180~~°	19'	14"
179°	79'	14"

→

179°	79'	14"
−90°	21'	3"
89°	58'	11"

Practical Problems

4.

120°	37'	14"
−38°	17'	6"

5.

98°	43'	21"
−18°	7'	43"

6.

75°	36'	0"
−22°	41'	28"

7.

123°	0'	23"
−90°	23'	45"

EXAMPLE 5: Multiplication

NOTE: Multiply each column separately.

A.

75°	36'	19"
		×3

→ B.

75°	36'	19"
		×3
		57"

C.

75°	36'	19"
		×3
	108'	57"

→ D.

75°	36'	19"
		×3
225°	108'	57"

E.

225°	108'	57"
+1	−60	
226°	48'	57"

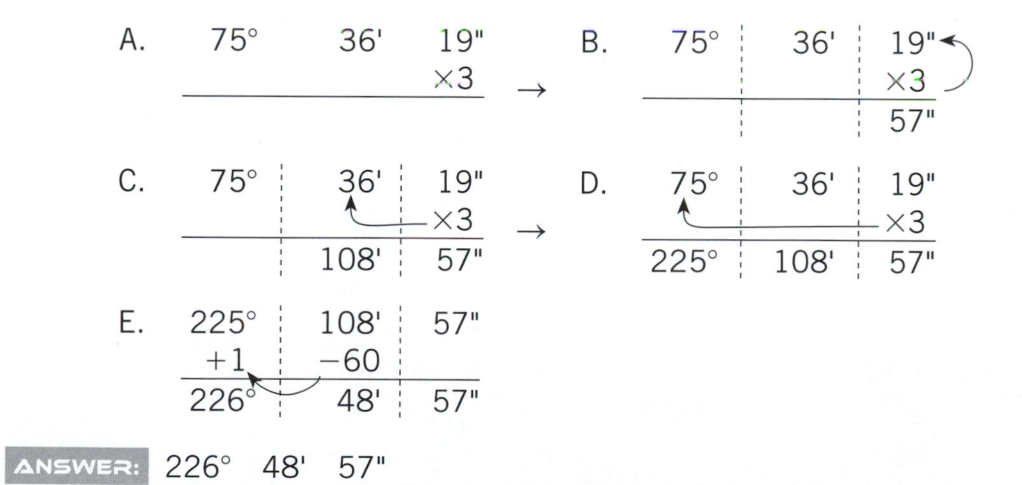

ANSWER: 226° 48' 57"

Practical Problems

8. 16° 48'
   × 4

9. 45° 6"
   ×4

10. 68° 41' 35"
    × 2

11. 110°
    ×3

EXAMPLE 6: **Division**

NOTE: Divide each column/unit separately.

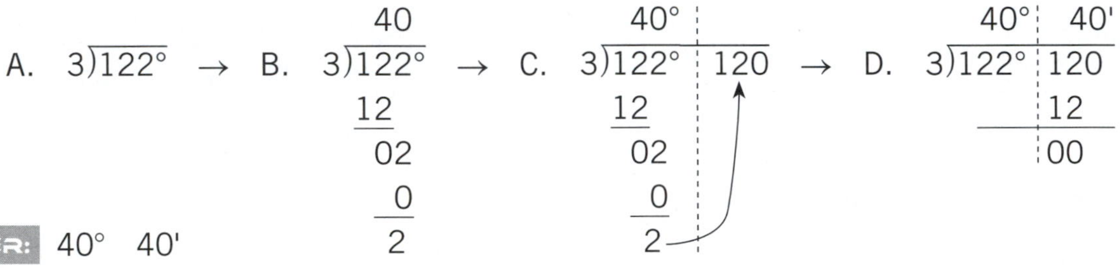

ANSWER: 40° 40'

EXAMPLE 7: **Division**

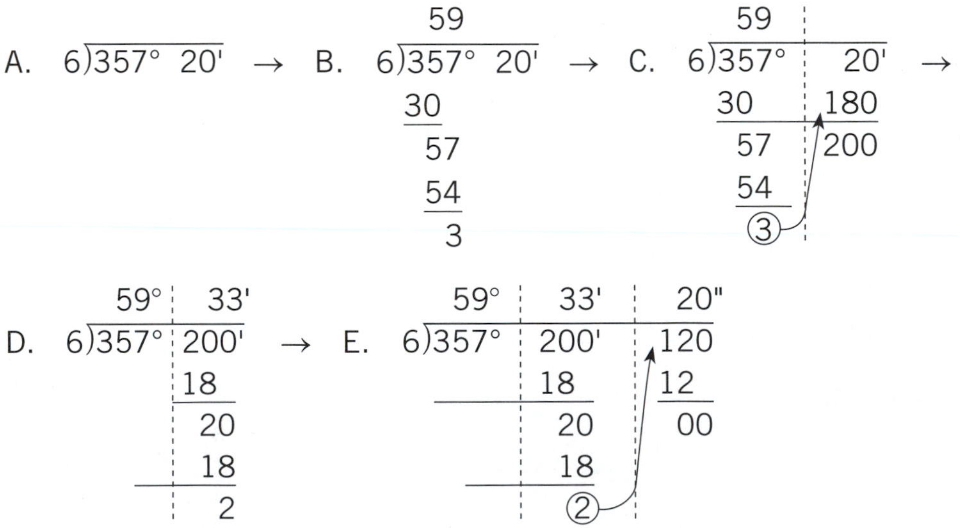

ANSWER: 59° 33' 20"

Practical Problems

12. $2\overline{)45^\circ}$

13. $\frac{1}{2} \times 275^\circ$

14. $3\overline{)273^\circ\ \ 22'}$

15. $\frac{1}{4} \times 275^\circ$

16. $2\overline{)139^\circ\ \ 17'\ \ 21''}$

UNIT 37

Protractors

Basic Principles

A protractor is an instrument in the form of a graduated semicircle showing degrees, used for drawing and measuring angles. See illustration A.

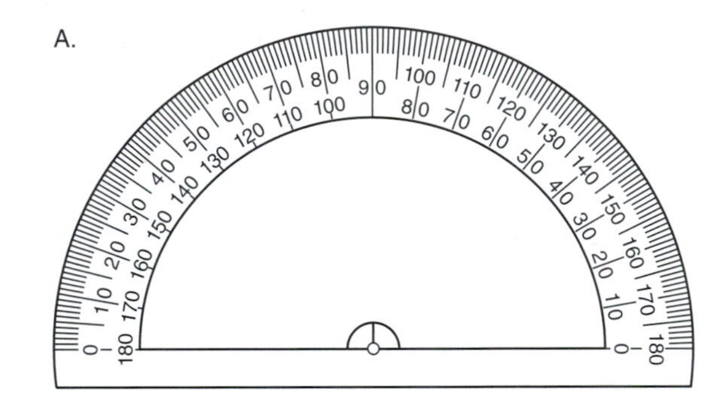

Using the Protractor to Draw a 40-Degree Angle

Draw a straight line.

A.

Place the protractor on the line with the 0° – 180° baseline in alignment. (Check your protractor; the bottom edge might not be the 0° – 180° baseline.)

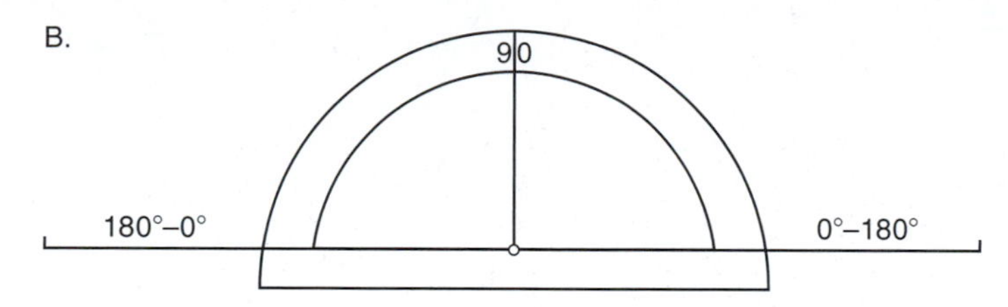

Place a mark in the center hole on the 0° – 180° baseline and a mark at the 40° position above the protractor.

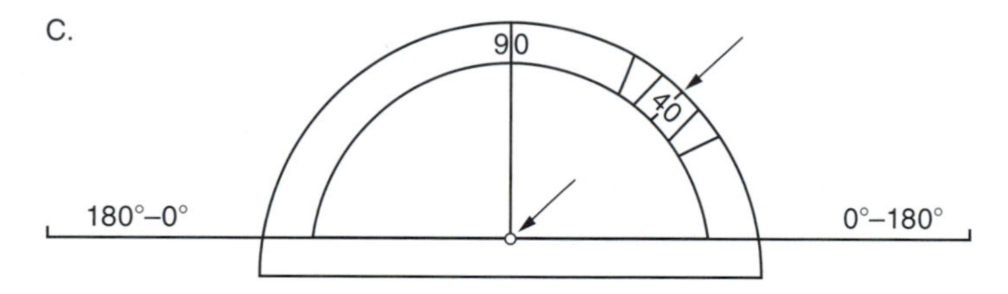

Remove the protractor, draw a straight line from the center hole mark, through the 40° mark, and extend.

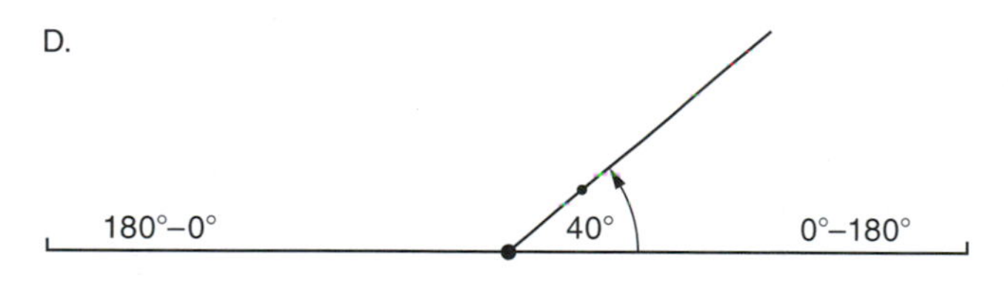

Practical Problems

1. Using a protractor, draw each angle.

   a. 45°

   b. 30°

   c. 90°

   d. 135°

   e. 22° 30'

2. Using a protractor, measure each angle. Extend lines as needed.

$\angle A =$ ____________

$\angle B =$ ____________

$\angle C =$ ____________

$\angle D =$ ____________

$\angle E =$ ____________

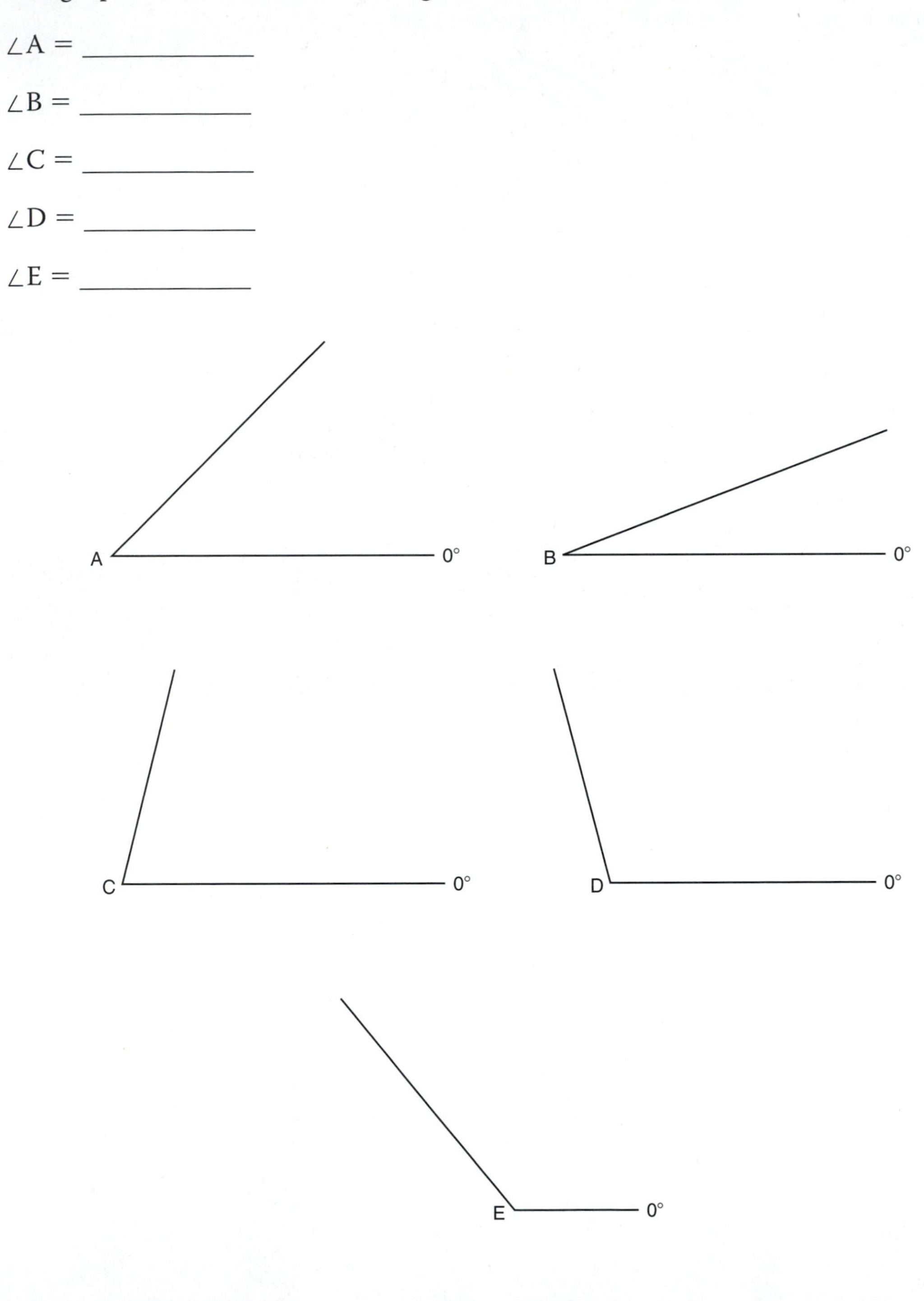

SECTION 8

BENDS, STRETCHOUTS, ECONOMICAL LAYOUT, AND TAKEOFFS

UNIT 38

Bends and Stretchouts of Angular Shapes

Basic Principles

As needed, flat plate can be bent to form angle, channel, and square or round pipe.

The length of the plate used changes in the bending process. The amount of change depends upon the bend angle, the bend radius, and the material thickness. It is called the "bend allowance."

In general, as plate is bent into the shape needed, the metal along the outside of the bend stretches and expands, while the metal inside the bend compresses. See Figures A and B.

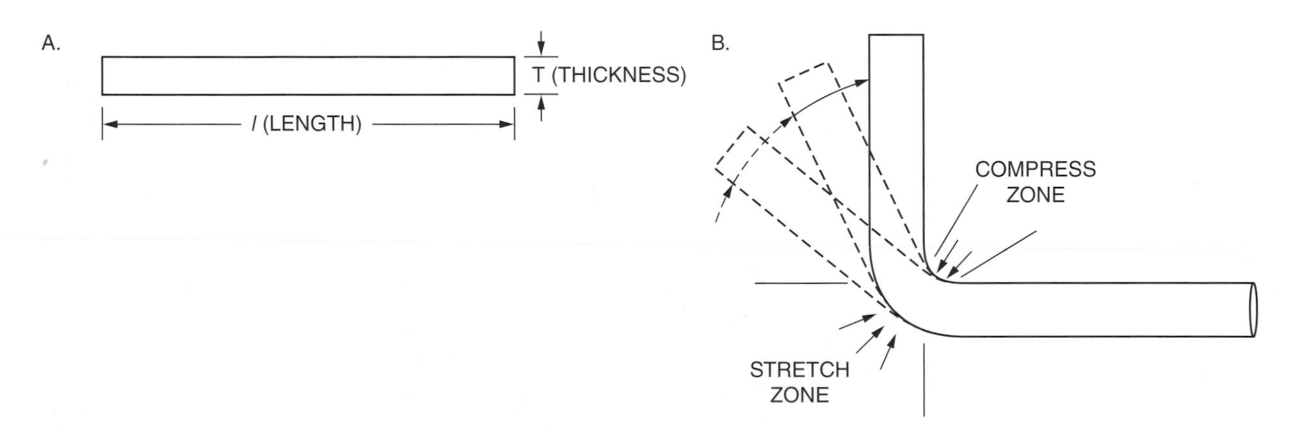

This change in size must be taken into consideration, and the appropriate bend allowance is usually found on charts used for that purpose.

However, the methods used in the next two units give appropriate calculations in determining the correct length of flat plate for bending into right angles (90°) and circular shapes.

Inside and Outside Bends

A bend is defined as inside when sides are given with I/S dimensions. See Figure C.

A bend is defined as outside when sides are given with O/S dimensions. See Figure D.

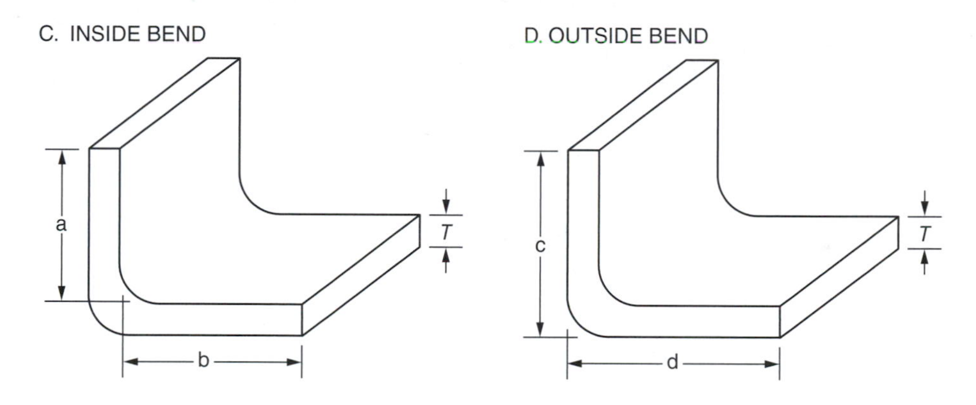

NOTE: To determine the correct length of plate needed to bend into a 90° angle, half the plate thickness (T) is either added or subtracted, per bend, to the calculations.

On inside bends, $\frac{1}{2}T$ is added to the side measurements given.

On outside bends, $\frac{1}{2}T$ is subtracted from the side measurements given.

EXAMPLE 1: Calculate the length of ½″ plate needed to bend a 90° angle with I/S leg measurements of a = 2.70″, and b = 2.70″.

NOTE: Formula

$$\ell = a + b + \frac{1}{2}T$$

$$= 2.70 + 2.70 + \frac{1}{2}(.50)$$

$$= 5.40 + .25$$

$$\ell = 5.65''$$

ANSWER: Length of plate = 5.65″

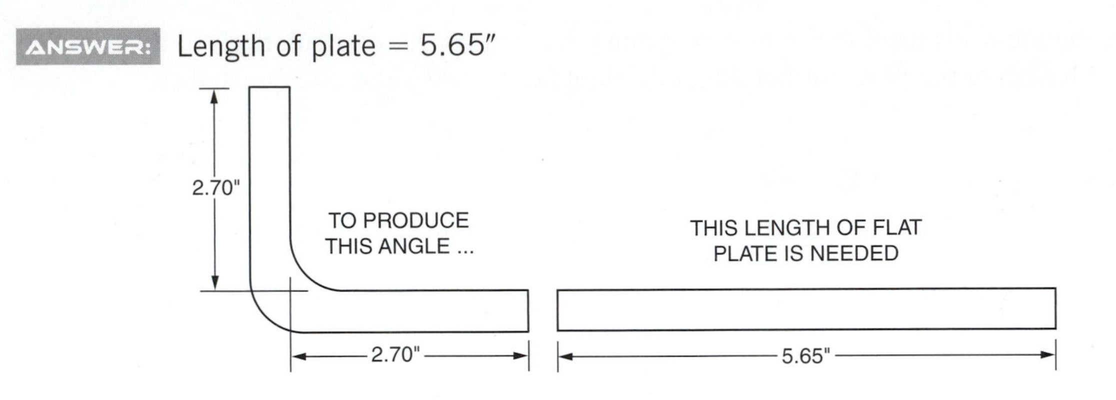

EXAMPLE 2: Square pipe with O/S wall measurements of 18.375″ is made from ⅛″ plate steel. Calculate the length of plate needed to manufacture this pipe. Three outside bends are used.

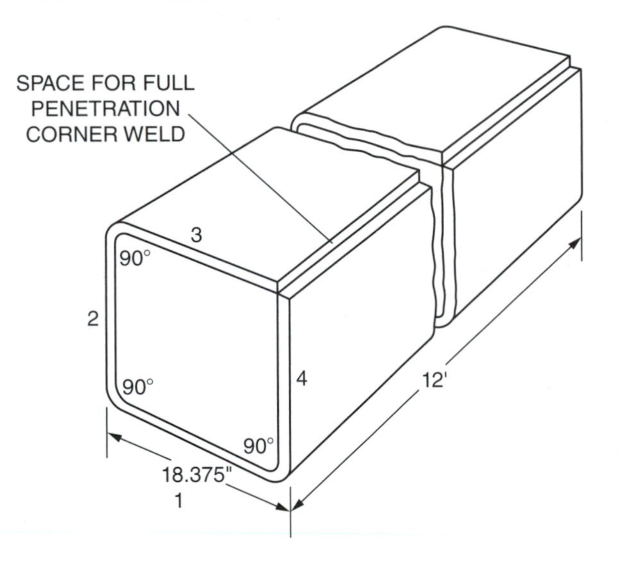

SOLUTION:

STEP 1: Add all wall lengths of steel together.

Wall 1: 18.375″
Wall 2: 18.375″
Wall 3: (18.375 − .125) = 18.250″
Wall 4: (18.375 − .125) = 18.250″

$$\begin{array}{r} 18.375 \\ 18.375 \\ 18.250 \\ +\ 18.250 \\ \hline 73.250'' \end{array}$$

STEP 2: Calculate ½-plate thickness. Then, multiply by the number of bends.

$$\frac{1}{2}T = .0625 \rightarrow \begin{array}{r} .0625 \\ \times\ 3 \\ \hline .1875'' \end{array} \text{ (there are 3 bends)}$$

STEP 3: Subtract this figure from the total length.

$$\begin{array}{r} 73.2500'' \\ -\ .1875'' \\ \hline 73.0625'' \end{array}$$

ANSWER: Length of plate needed = 73.0625″

Practical Problems

1. This machinery cover is of a square-welded, two-piece, 0.64 cm steel plate design. Two 90° outside bends are required for preparation. Find the length of each piece used for the weldment. ____________

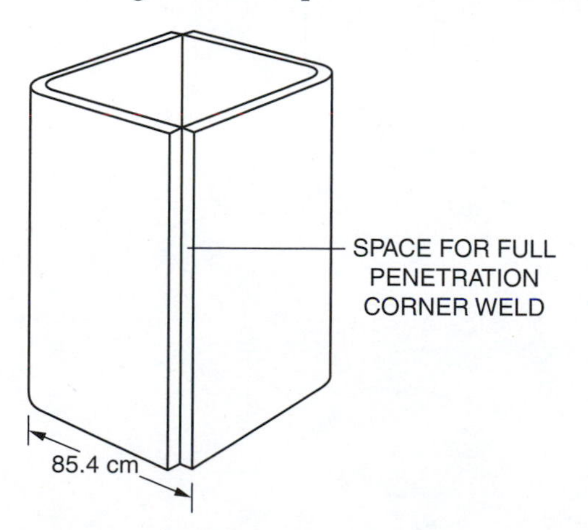

2. Two outside 90° bends are used to shape a section of transfer gutter. Find the length of 1⁄16″ plate needed for this order of 6′ gutter. _______________

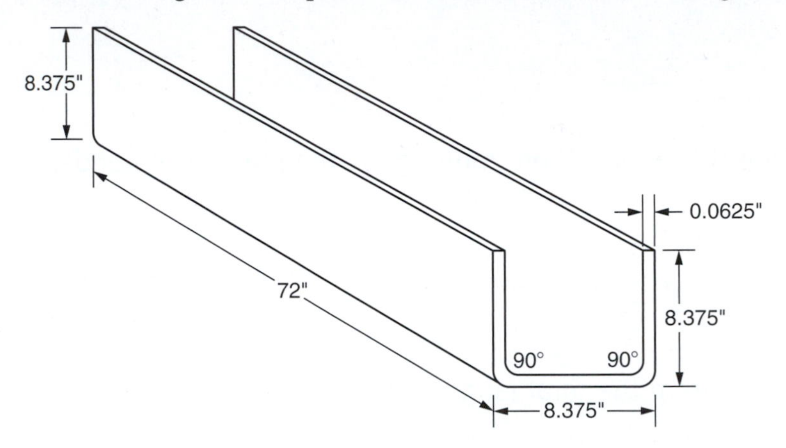

3. Four inside 90° bends are used in this gasoline tank. Find the size of 1⁄8″ plate steel used to complete the tank body. Make no allowance for a weld gap. _______________

NOTE: Answer should show size: length and width.

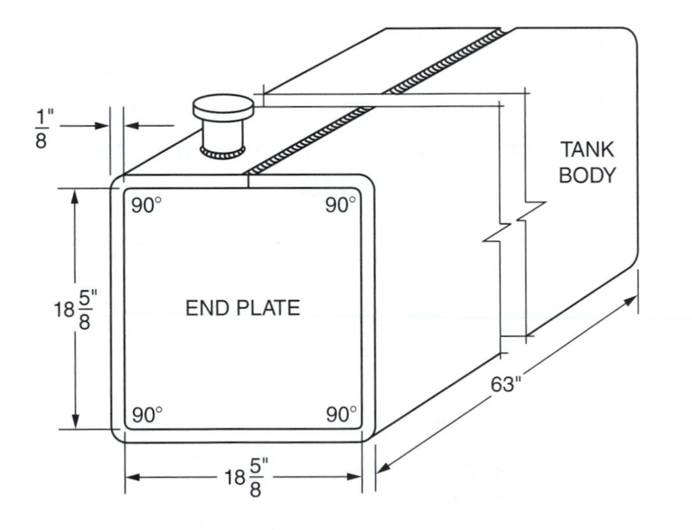

4. Forty-three open-end welded steel storage boxes are welded from 0.48-cm plate steel. The plate thickness is enlarged to show detail. Find the size of plate used for one storage box. ____________

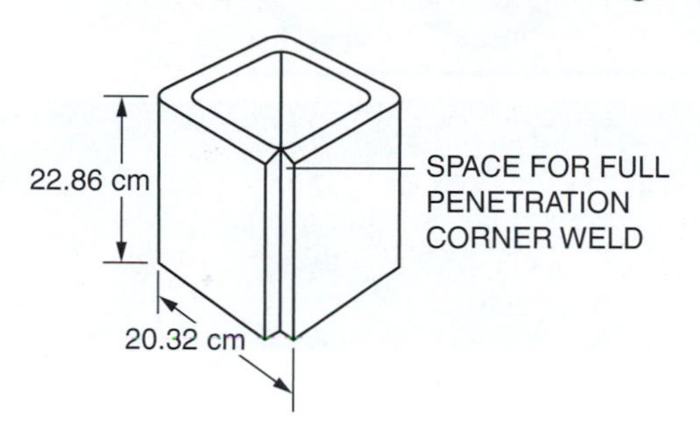

5. Four 90° outside bends are required for the tank shown. Find the length, in feet, of the steel piece needed to bend the main body of the tank. The material thickness is ⅛″. Make no allowance for a weld gap. ____________

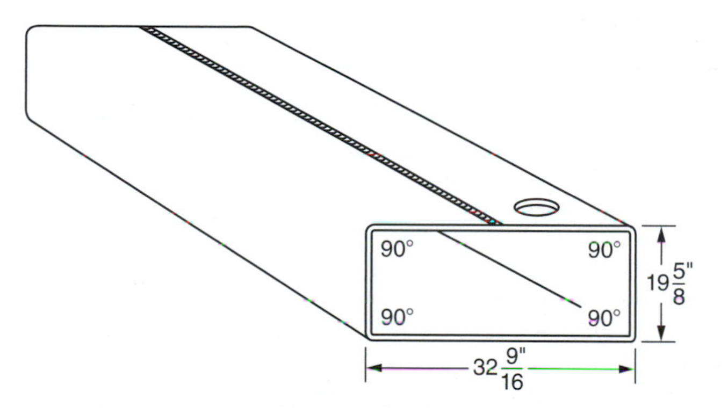

6. A welded high-pressure hydraulic tank is shown. Find the size of the 0.635-cm sheet of steel plate used to construct the tank. Outside bends are used. ____________

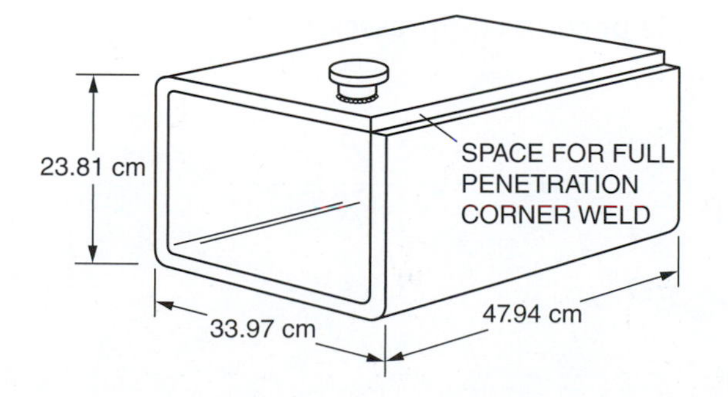

UNIT 39

Bends and Stretchouts of Circular and Semicircular Shapes

Basic Principles

When measuring flat plate that is to be bent and formed into round pipe, there is both an inside diameter/circumference and an outside diameter/circumference to take into consideration. This is due to plate thickness and metal changes during the bending process.

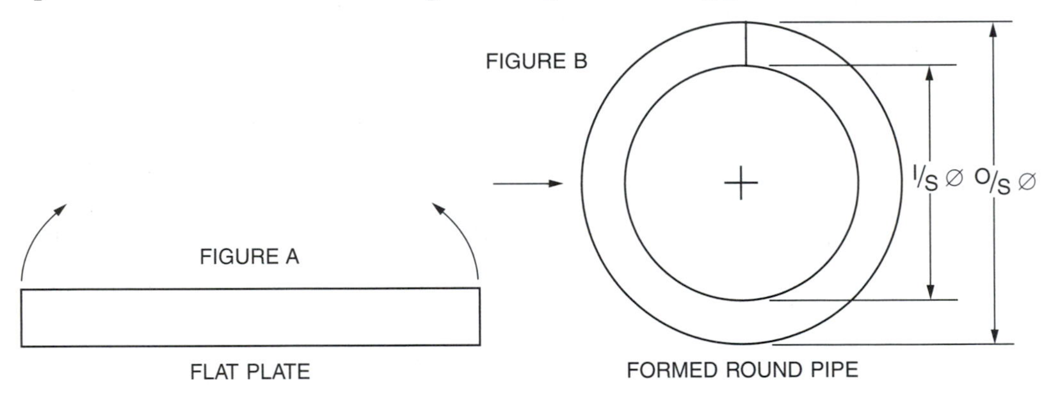

As in forming angle from plate (see previous unit), during the circular bending process, the outside portion of the metal stretches, while the inside portion compresses.

Reminder

The formula for the circumference of a circle is

$$C = \pi D$$

To properly calculate the length of plate needed (circumference) to form a circular pipe, the average of the two diameters is used.

There are several methods available to calculate the average diameter:

1. add both diameters together; divide by 2;
2. subtract a pipe-wall thickness from the O/S diameter; or
3. add a pipe-wall thickness to the I/S diameter.

EXAMPLE: Find the size of the ½″ plate needed to form this circular drainpipe.

SOLUTION: See Figure B for clarity.

STEP 1: Determine the average diameter.

Method 1: Averaging

O/S diameter = 18″ →
I/S diameter = +17″
35″

$$\begin{array}{r} 17.5'' \\ 2\overline{)35.0} \\ \underline{2} \\ 15 \\ \underline{14} \\ 10 \\ \underline{10} \end{array}$$

Average diameter = 17.5″

Method 2: O/S diameter minus wall thickness

$$\begin{array}{r} 18'' \\ -\frac{1}{2}'' \\ \hline \end{array} \rightarrow \begin{array}{r} 17\frac{2}{2} \\ -\frac{1}{2} \\ \hline 17\frac{1}{2}'' \end{array}$$

Average diameter = 17½″

Method 3: I/S diameter plus wall thickness

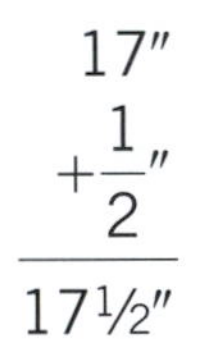

Average diameter = 17½″

NOTE: Each method produces the same answer.

STEP 2:

$C = \pi D$
$C = 3.14(17.5'')$
$C = 54.95''$

ANSWER: Size of ½″ plate needed = 54.95″ × 72″

Practical Problems

1. Find the size of the ¼″ plate needed to construct this semicircular ventilation section. The average diameter is 19¾₁₆″. ______________

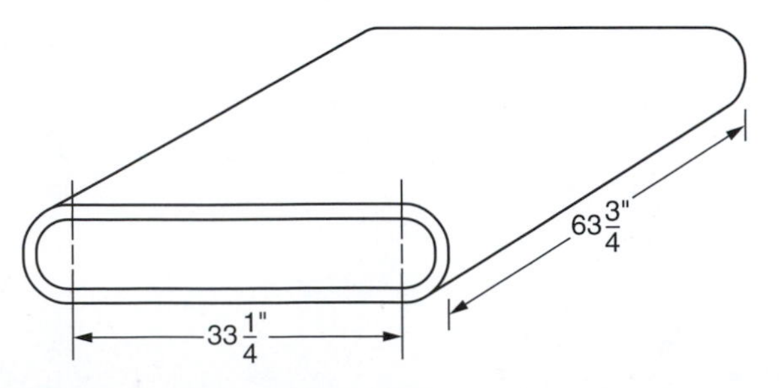

2. This hydraulic ram cylinder shown below is rolled from 1.5875-cm thick metal. Find the size of the plate steel needed to construct the cylinder. __________

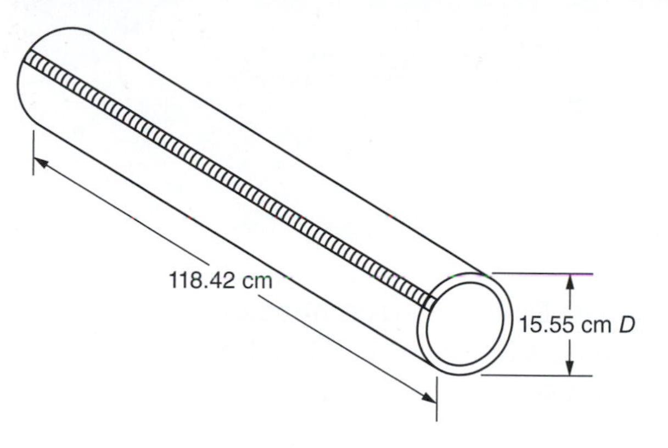

3. This semicircular-sided tank is rolled from 3⁄16″ plate. The average diameter is 19 11⁄16″.

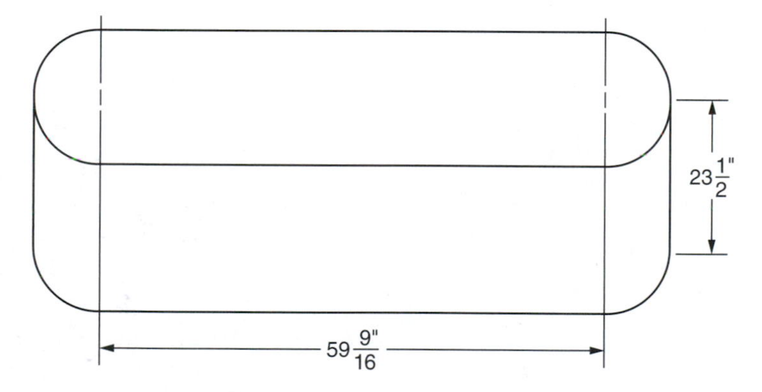

a. Find the length of 3⁄16″ plate needed to roll the tank. a. __________

b. The bottom of the tank is cut from a rectangular piece of 3⁄16″ plate. Find the width of the plate. b. __________

c. Find the length of plate needed for the bottom of the tank. c. __________

4. This electrode holding-tube has an average radius of 6.66 cm. Find the size of the 17.46-cm single sheet of ¼″ plate metal needed to construct 10 tubes. __________

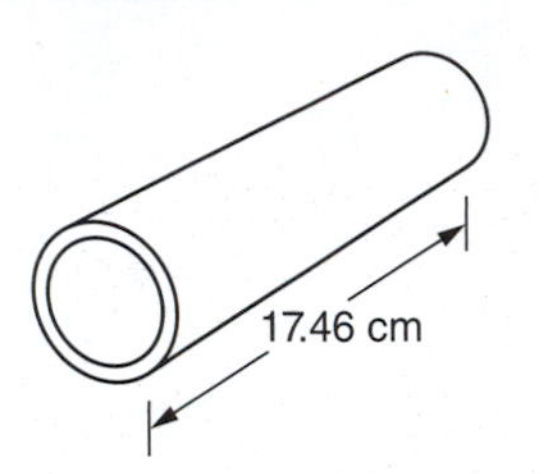

5. A branch header is constructed as shown. Find the stretchout of the material needed to construct the two pieces of the branch header from 3⁄16″ steel plate. Make no allowances for weld gaps or seams.

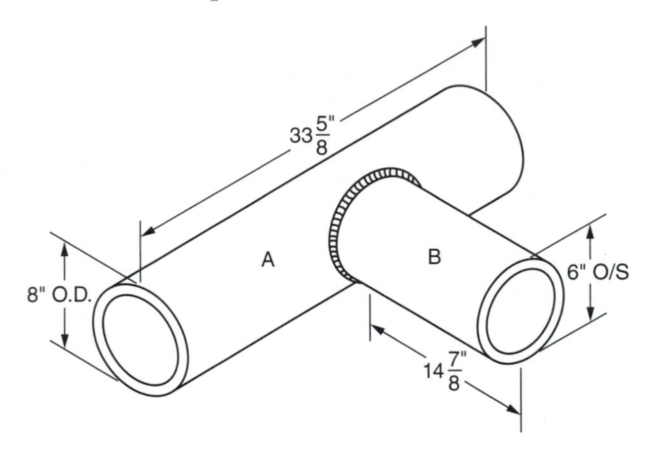

a. Branch header part A — a. __________

b. Branch header part B — b. __________

6. The average diameter of this storage tank is 9 feet 6½″. Find the number of ⅛″ sheet metal plates needed to complete the cylindrical side of this storage tank. The sheet size available is ⅛″ × 48″ × 96″. __________

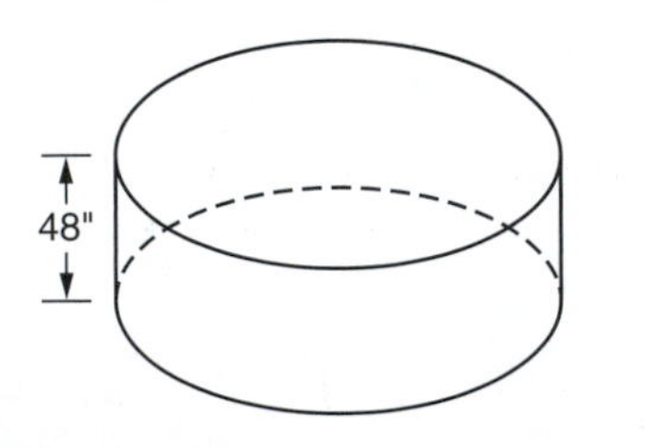

UNIT 40

Economical Layouts of Rectangular Plates

Basic Principles

There are two methods to find the most economical layout of a steel plate: answers should show the maximum number of pieces that can be obtained.

1. Make two sketches, laying out the length and width of the pieces to be cut both possible ways.
2. Mathematical calculations.

EXAMPLE: What is the most number of coupons 3″ × 7″ that can be cut from a plate 8″ × 22″? Disregard width of the cut.

SOLUTION:

Layout 1 Sketch

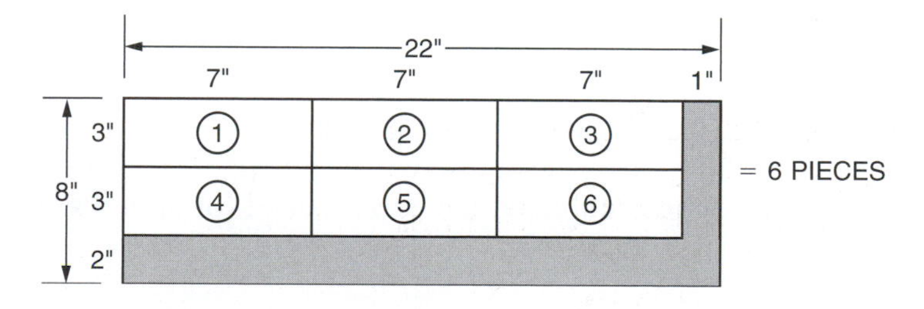

NOTE: Disregard fractional portions or remainders: only whole pieces are counted.

Mathematical Steps

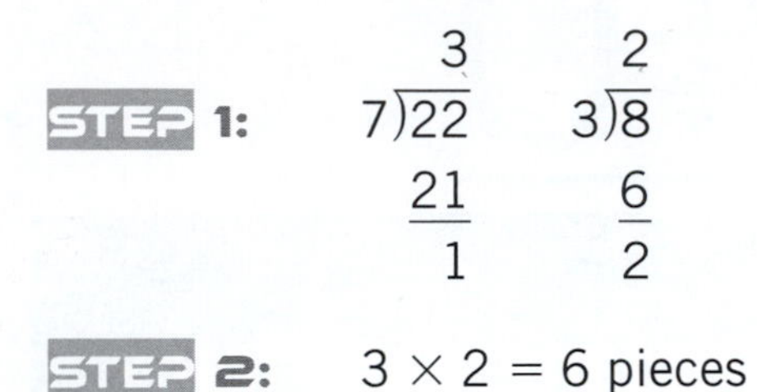

STEP 2: $3 \times 2 = 6$ pieces

Layout 2 Sketch

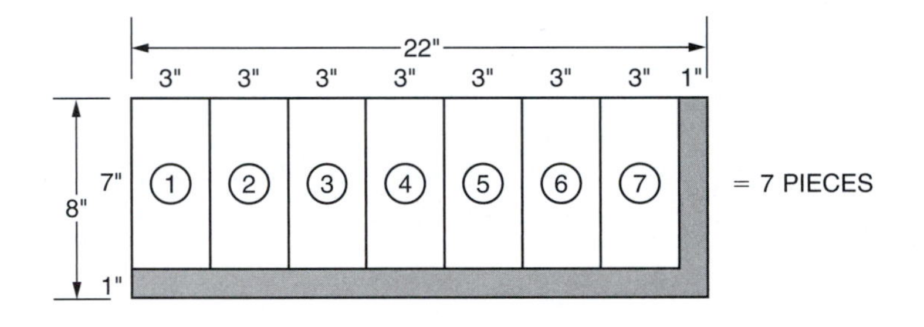

Mathematically

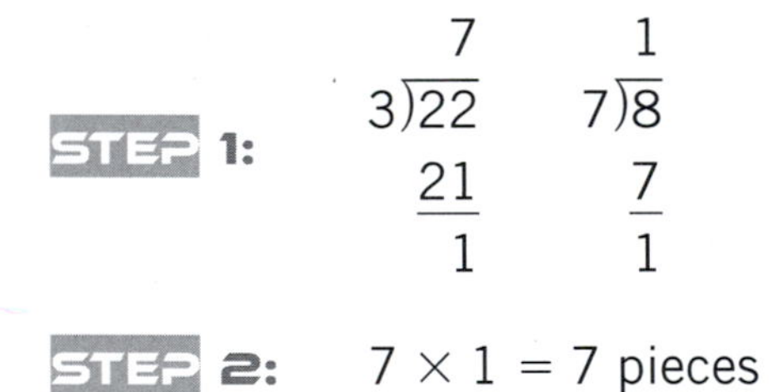

STEP 2: $7 \times 1 = 7$ pieces

ANSWER: 7 pieces. Layout 2 produced the most pieces.

The following is an additional type of table which can be used for speed and simplification:

STEP 1:

$22 \div 7 = 3$ $\quad$ $8 \div 7 = 1$

$22 \div 3 = 7$ $\quad$ $8 \div 3 = 2$

 Cross multiply

$22 \div 7 = ③ \quad 8 \div 7 = ① \longrightarrow 7 \times 1 = ⑦$

$22 \div 3 = ⑦ \quad 8 \div 3 = ② \longrightarrow 3 \times 2 = 6$

The table also shows that 7 pieces is the answer.

Practical Problems

Answers should show the maximum number of pieces that can be obtained. Disregard waste caused by the width of the cuts unless noted.

1. A weld shop supplies 104 shaft blanks, each 4″ wide and 5″ long. How many can be cut from the piece of plate shown (a)? How many can be cut with a .25 kerf width (b)?

   a. ______________

   b. ______________

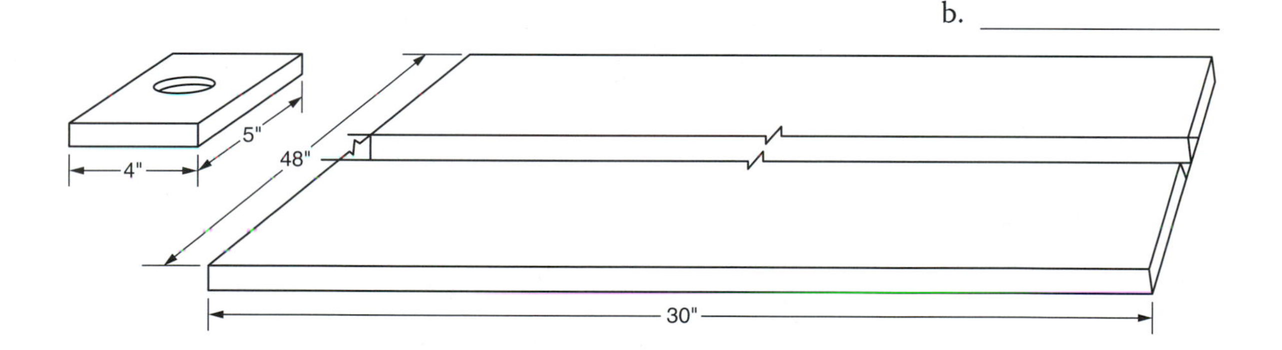

2. How many 12.7-cm by 15.24-cm plates can be cut from this plate? ______________

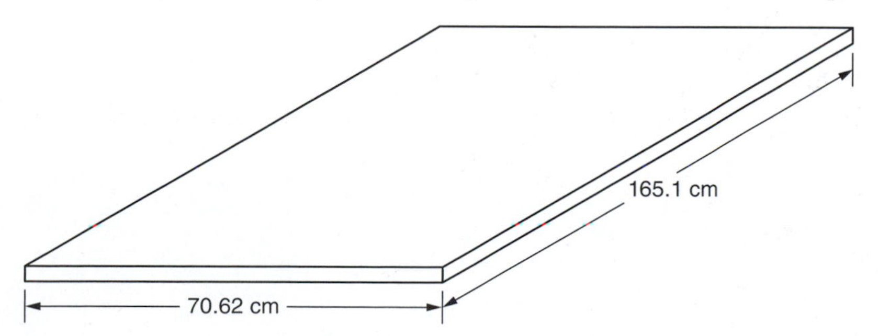

3. These pulley brackets are cut from a 1″ thick plate of steel that is 6½″ × 48½″.

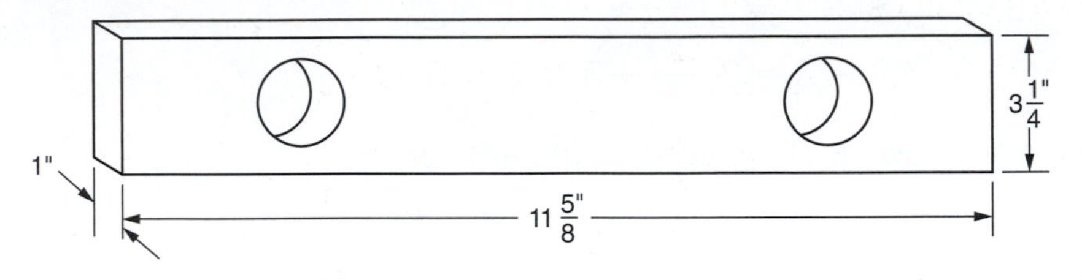

a. How many of these brackets can be cut? a. ______________

b. What is the size of the material remaining after the brackets are cut in the most economical way? b. ______________

4. Column baseplates of the size shown are cut and drilled. The baseplates are cut from a steel plate with the dimensions of 84.750″ by 89.500″. Width of the cut is 0.125″. How many baseplates can be cut? ______________

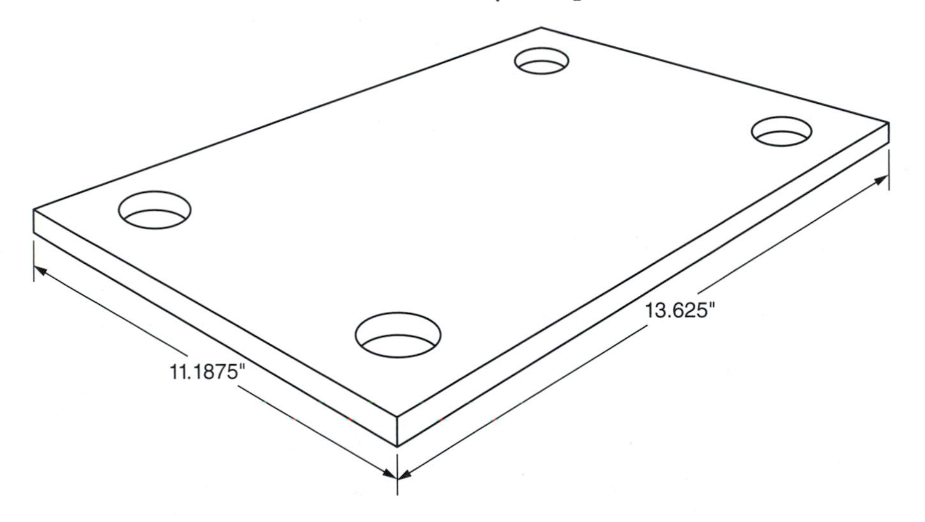

5. These angle brackets are cut and welded to finish a construction job. How many of the brackets can be obtained from a steel plate that is 60″ wide by 90½″ long? Width of cut is 0.25″. __________

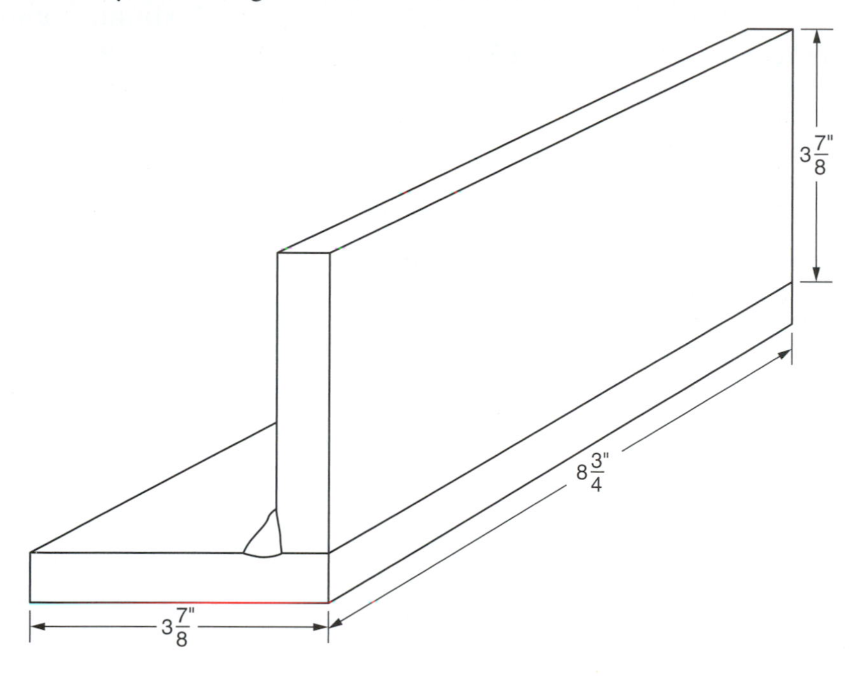

6. Storage bin sides are 20.32 cm by 27.94 cm. Three sides are welded together for each bin. How many bins can be made from the plate of steel shown? __________

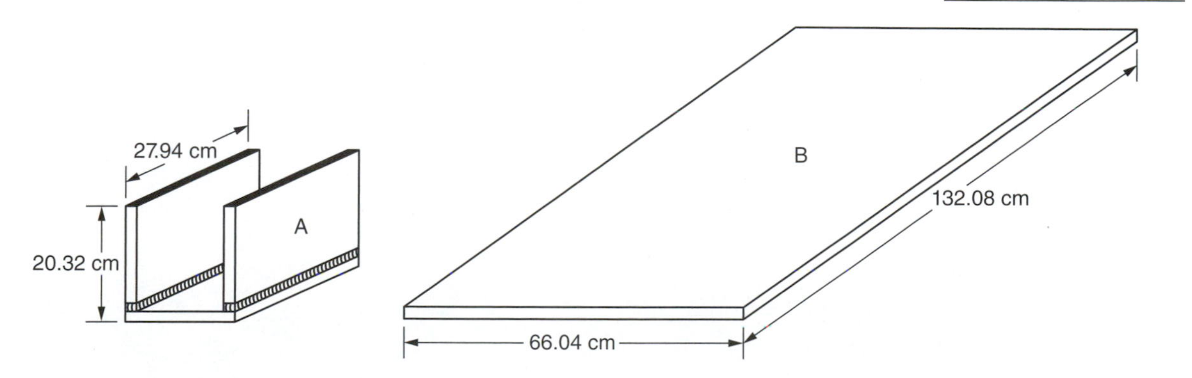

Find the maximum number of pieces of the given sizes that can be cut from the indicated sheets of steel.

	Dimensions of Piece	Sheet Size	Maximum Number of Pieces
7.	7 in × 8 in	36 in × 71 in	________
8.	8 in × 11 in	48 in × 120 in	________
9.	25.4 cm × 33.02 cm	152.4 cm × 304.8 cm	________
10.	21 in × 27 in	3 ft 9 in × 5 ft	________
11. a.	5.08 cm × 15.24 cm	33.02 cm × 137.16 cm	a. ________
b.	Calculate Question 11 with .15 cm kerf.		b. ________

UNIT 41

Economic Layout of Odd-Shaped Pieces; Takeoffs

Basic Principles

Sketches of different arrangements may be helpful. Disregard waste caused by the width of the cuts unless noted. Layout of curved or odd-shaped pieces may work best using drawings instead of the mathematical method.

EXAMPLE: A piece of scrap metal of the shape shown below is cut into 10″ radius circles. How many circles can be cut from the material if the kerf width is disregarded?

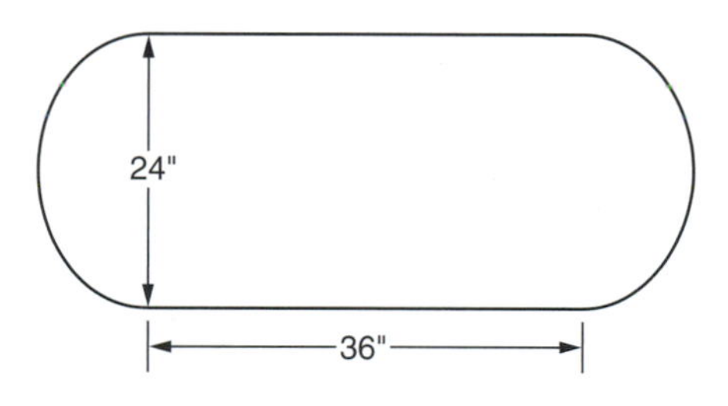

SOLUTION: Layout

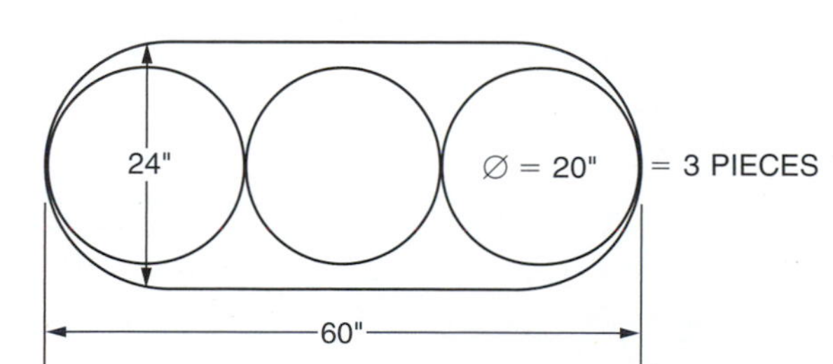

Mathematically

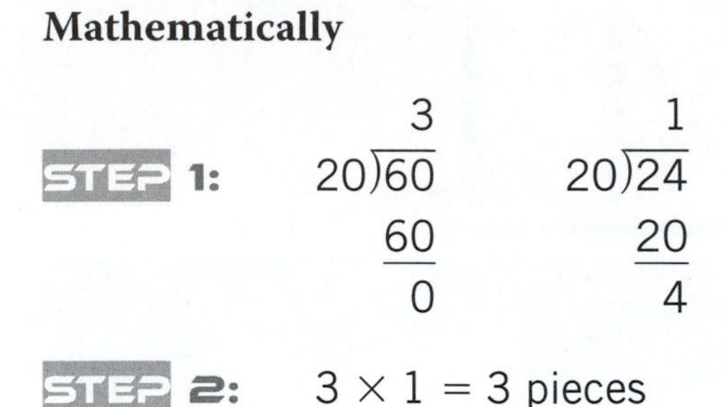

STEP 2: $3 \times 1 = 3$ pieces

Practical Problems

1. a. How many 13″ diameter circles can be cut from the scrap metal used in the previous example? __________

   b. From this size scrap, how many 20″ diameter circles can be cut if the kerf has a 3⁄16″ width? __________

2. How many pieces of sheet metal 3⁄4″ wide and 60″ long can be cut from this sheet? __________

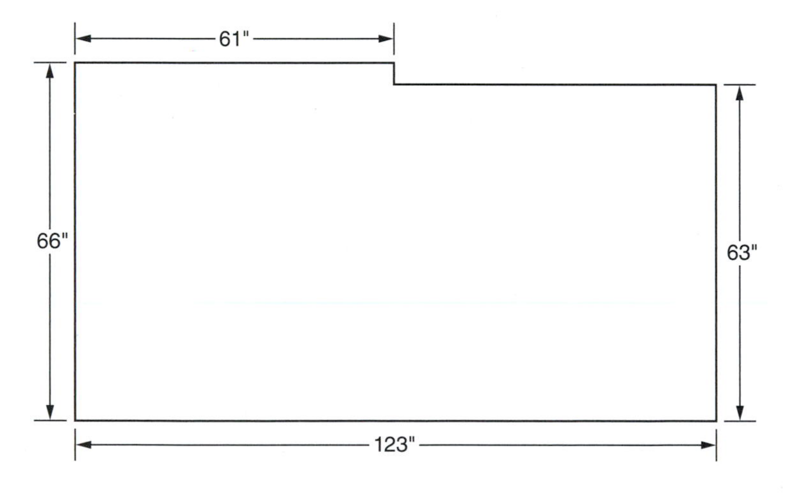

3. Fourteen sections of 2″ pipe of the length shown are used to construct a framework.

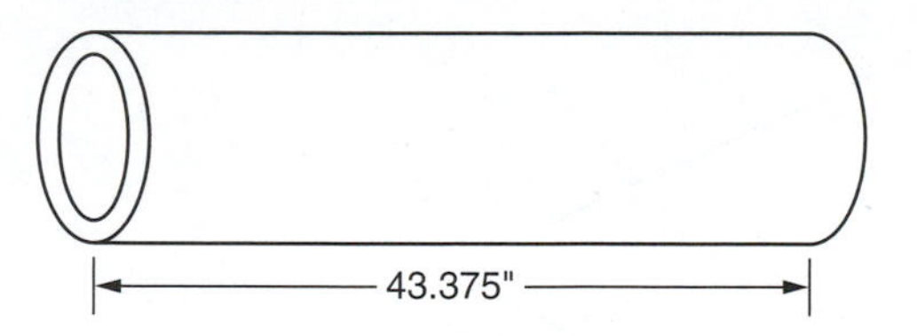

a. How many standard 21′ lengths of pipe must be used to cut the 14 sections? a. ______________

b. What percent of the pipe used is wasted after cutting? Round the answer to the nearest hundredth percent. b. ______________

4. Thirty-one column gussets are cut from a steel plate. Find the dimensions of the smallest square plate from which all 31 gussets can be cut. ______________

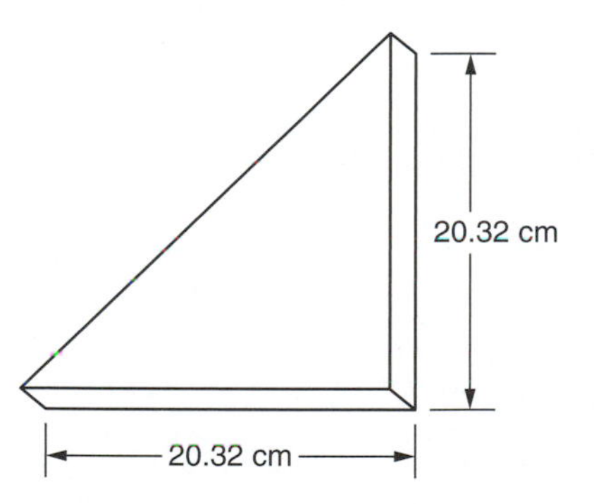

5. This circular blank is used to make sprocket drives. How many sprocket drive blanks can be cut from a plate of steel having the dimensions of 44″ × 44″? ______________

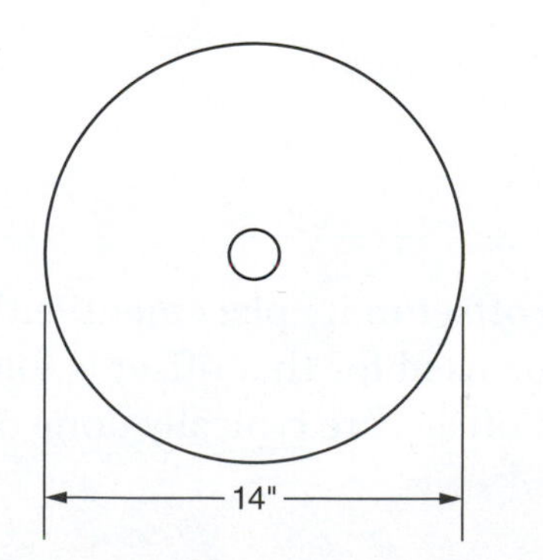

6. Gussets are cut from the material shown. How many gussets can be cut from one sheet? ____________

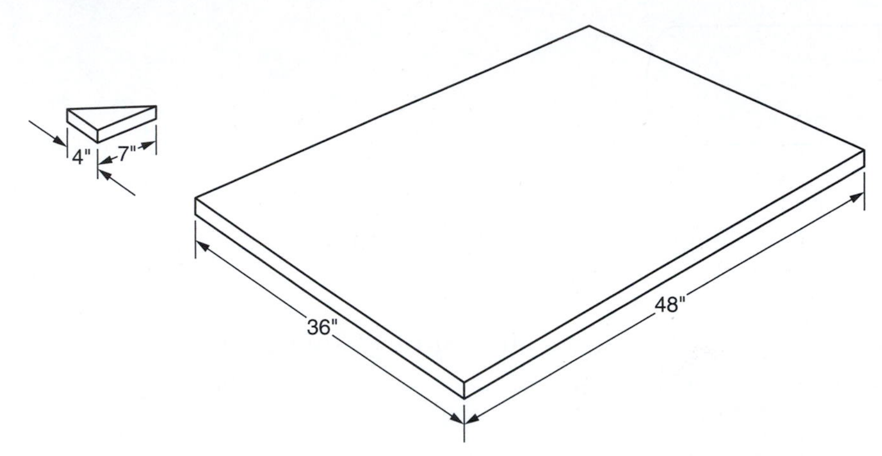

7. How many 4″ × 3″ rectangular test plates can be cut from the piece of scrap? ____________

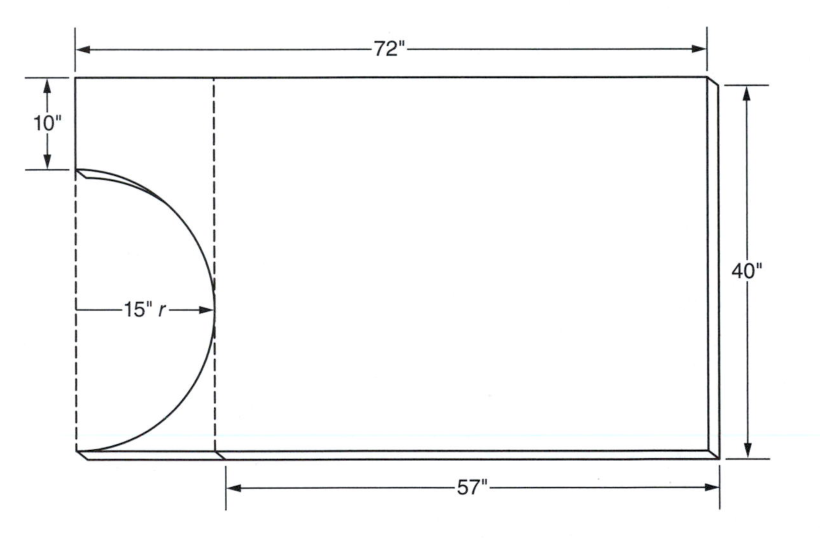

Basic Principles

"Take-offs" is a term used in pipefitting. When a pipe has to be offset in its placement, either to avoid an obstacle or for design purposes, the length of the pipe used for that offset is found by subtracting (taking off) measurements per fitting. Fittings and offsets are typically done on a 90-degree angle or a 45-degree angle. See Figures A and B.

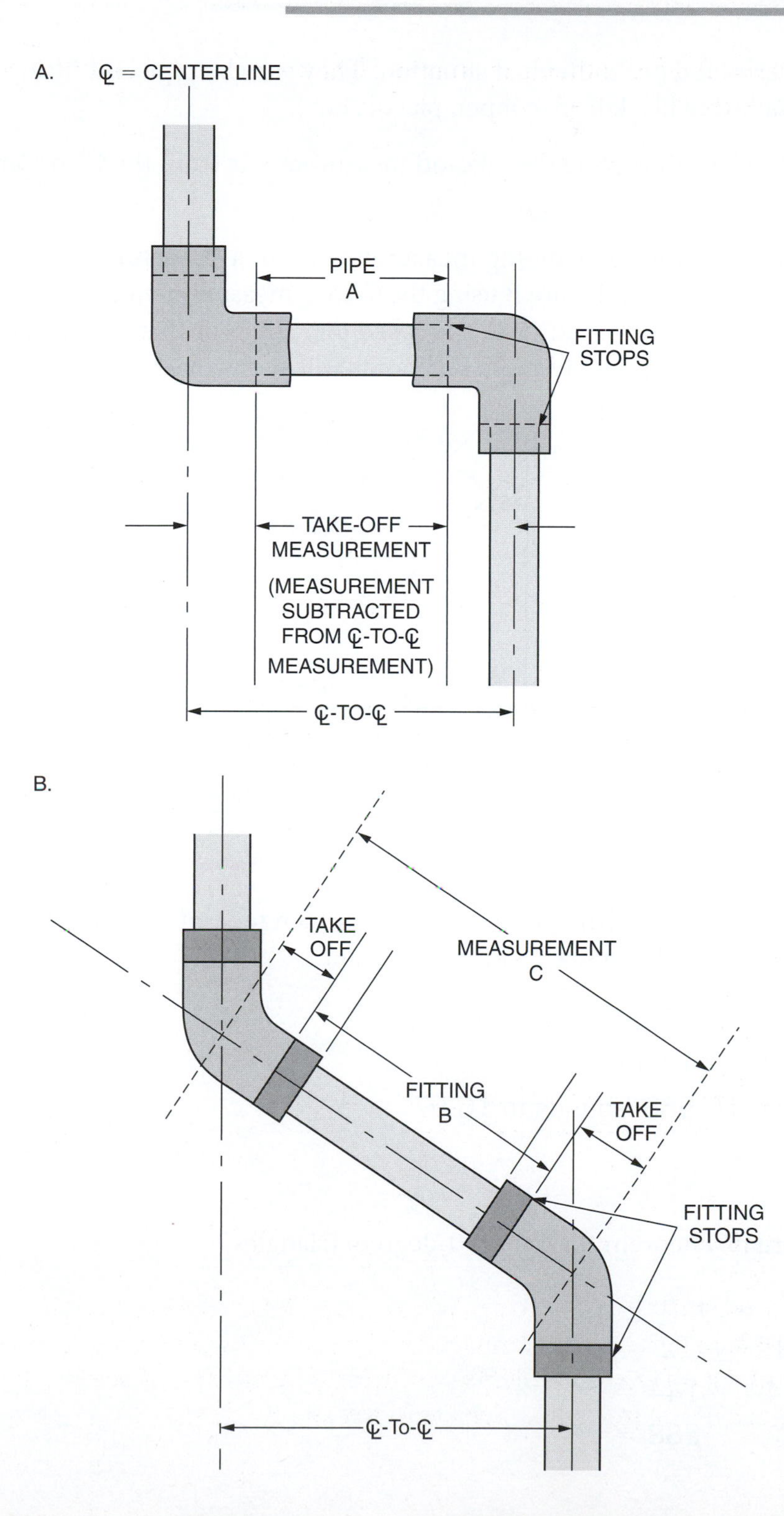
A.
℄ = CENTER LINE
PIPE
A
FITTING
STOPS
TAKE-OFF
MEASUREMENT
(MEASUREMENT
SUBTRACTED
FROM ℄-TO-℄
MEASUREMENT)
℄-TO-℄
B.
TAKE
OFF
MEASUREMENT
C
FITTING
B
TAKE
OFF
FITTING
STOPS
℄-To-℄

Take-offs are determined per individual situation. They vary by pipe and fitting size, and by the material used (malleable black iron, copper, plastic, etc.).

Pipe A is calculated by subtracting the take-off measurements from the ℄-to-℄ measurement, as seen in Figure A.

There are two methods for determining measurement c in a 45-degree offset. Develop a right triangle from the illustration in Figure B using the ℄-to-℄ measurement as the base (see Figure C). This measurement is called the "run". Side b is called the "rise". For this example, we'll assign 12″ to side *a* and 12″ to side *b* of the right triangle. Blueprints will give actual measurements for fieldwork.

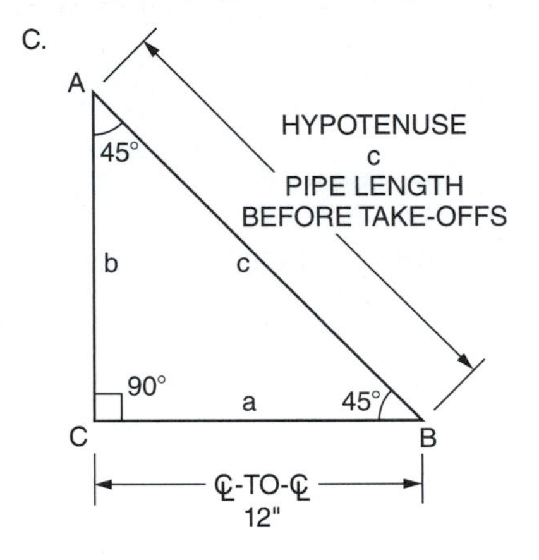

Method 1

To determine measurement *c* (the pipe length from which take-offs will be subtracted), multiply the run length by 1.414. The figure 1.414 is a constant.

$$\begin{aligned} c &= 1.414\,(a) \\ &= 1.414(12'') \\ &= 16.968'' \text{ rounds to } 16.97'' \end{aligned}$$

Method 2

Use the Pythagorean Theorem for right (90-degree) triangles.

$$\begin{aligned} a^2 + b^2 &= c^2 \\ 12^2 + 12^2 &= c^2 \\ 144 + 144 &= c^2 \\ 288 &= c^2 \end{aligned}$$

To find c, determine the square root of 288.

$$\sqrt{288} = \sqrt{c^2}$$
$$\sqrt{288} = c$$

Enter 288 in your calculator and press the $\sqrt{\ }$ button.

$$\sqrt{288} = 16.97$$
$$c = 16.97''$$

Practical Problems

8. Determine the length of pipe needed in a 90-degree offset, ℄-to-℄ of 35″ and the following take-offs per fitting side:

   a. .75″ a. ______

   b. 2.35″ b. ______

   c. .38″ c. ______

In Problems 2 and 3, the rise measurement will be the same as the run.

9. Using the constant figure in method 1, determine the length c of a 45-degree offset with the following figures for the run:

   a. 73″ a. ______

   b. 2′ 8″ (answer in feet and inches) b. ______

   c. 24.5″ c. ______

10. Using the Phythagorean Theorem, calculate the length c of a 45-degree offset with the following run figures:

    a. 24″ a. ______

    b. 3′ 10″ (answer in feet and inches) b. ______

    c. 62.75″ c. ______

APPENDIX

Section 1

Denominate Numbers

Denominate numbers are numbers that include units of measurement from the same group, such as yards, feet, and inches, or hours, minutes, and seconds. The units of measurement are arranged from the largest at the left to the smallest at the right.

EXAMPLES: 6 yards, 2 feet, 8 inches 12 hours, 3 minutes, 10 seconds

I. Basic Operations: See Unit 36, Angular Measurement

Measurements that are equal can be expressed in different terms.

EXAMPLES:

12 inches = 1 foot

100 centimeters = 1 meter

60 minutes = 1 degree, or 1 hour

1 inch = 2.54 centimeters

II. Equivalent Measures

All basic operations of arithmetic can be performed on denominate numbers: addition, subtraction, multiplication, and division. Including final corrections, all operations except division are begun with the smallest unit on the right. To perform operations on denominate numbers, keep like-units in the same column separate from other columns.

Section 2

Equivalents

English Relationships

English Length Measure

1 foot (ft)	=	12 inches (in)
1 yard (yd)	=	3 feet (ft)
1 mile (mi)	=	1,760 yards (yd)
1 mile (mi)	=	5,280 feet (ft)

English Area Measure

1 square yard (sq yd)	=	9 square feet (ft^2)
1 square foot (ft^2)	=	144 square inches (in^2)
1 square mile (sq mi)	=	640 acres
1 acre	=	43,560 square feet (ft^2)

English Volume Measure for Solids

1 cubic yard (cu yd)	=	27 cubic feet (cu ft)
1 cubic foot (ft^3)	=	1,728 cubic inches (in^3)

English Volume Measure for Fluids

1 quart (qt)	=	2 pints (pt)
1 gallon (gal)	=	4 quarts (qt)

English Volume Measure Equivalents

1 gallon (gal)	=	0.133681 cubic foot (cu ft)
1 gallon (gal)	=	231 cubic inches (cu in)

DECIMAL AND METRIC EQUIVALENTS OF FRACTIONS OF AN INCH

FRACTION	1/32nds	1/64ths	DECIMAL	MILLIMETERS
		1	0.015625	0.3968
	1	2	0.03125	0.7937
		3	0.046875	1.1906
1/16	2	4	0.0625	1.5875
		5	0.078125	1.9843
	3	6	0.09375	1.3812
		7	0.109375	2.7780
1/8	4	8	0.125	3.1749
		9	0.140625	3.5718
	5	10	0.15625	3.9686
		11	0.171875	4.3655
3/16	6	12	0.1875	4.7624
		13	0.203125	5.1592
	7	14	0.21875	5.5561
		15	0.234375	5.9530
1/4	8	16	0.250	6.3498
		17	0.265625	6.7467
	9	18	0.28125	7.1436
		19	0.296875	7.5404
5/16	10	20	0.3125	7.9373
		21	0.328125	8.3341
	11	22	0.34375	8.7310
		23	0.359375	9.1279
3/8	12	24	0.375	9.5240
		25	0.390625	9.9216
	13	26	0.40625	10.3185
		27	0.421875	10.7154
7/16	14	28	0.4375	11.1122
		29	0.453124	11.5091
	15	30	0.46875	11.9060
		31	0.484375	12.3029
1/2	16	32	0.500	12.6997

(Continued)

DECIMAL AND METRIC EQUIVALENTS OF FRACTIONS OF AN INCH

FRACTION	1/32nds	1/64ths	DECIMAL	MILLIMETERS
		33	0.515625	13.0966
	17	34	0.53125	13.4934
		35	0.546875	13.8903
9/16	18	36	0.5625	14.2872
		37	0.578125	14.6841
	19	38	0.59375	15.0809
		39	0.609375	15.4778
5/8	20	40	0.625	15.8747
		41	0.640625	16.2715
	21	42	0.65625	16.6684
		43	0.671875	17.0653
11/16	22	44	0.6875	17.4621
		45	0.703125	17.8590
	23	46	0.71875	18.2559
		47	0.734375	18.6527
3/4	24	48	0.750	19.0496
		49	0.765625	19.4465
	25	50	0.78125	19.8433
		51	0.796875	20.2402
13/16	26	52	0.8125	20.6371
		53	0.828125	21.0339
	27	54	0.84375	21.4308
		55	0.859375	21.8277
7/8	28	56	0.875	22.2245
		57	0.890625	22.6214
	29	58	0.90625	23.0183
		59	0.921875	23.4151
5/16	30	60	0.9375	23.8120
		61	0.953125	24.2089
	31	62	0.96875	24.6057
		63	0.984375	25.0026
1	32	64	1.000	25.3995

SI Metrics Style Guide

SI metrics is derived from the French name Système International d'Unites. The metric unit names are already in accepted practice. SI metrics attempts to standardize the names and usages so that students of metrics will have universal knowledge of the application of terms, symbols, and units.

The English system of mathematics (used in the United States) has always had many units in its weights and measures tables that were not applied to everyday use. For example, the pole, perch, furlong, peck, and scruple are not used often. These measurements, however, are used to form other measurements and it has been necessary to include the measurements in the tables. Including these measurements aids in the understanding of the orderly sequence of measurements greater or smaller than the less frequently used units.

The metric system also has units that are not used in everyday application. SI metrics, however, places an emphasis on the most frequently used units.

In using the metric system and writing its symbols, certain guidelines are followed. For the students' reference, some of the guidelines are listed.

1. In using the symbols for metric units, the first letter is capitalized only if it is derived from the name of a person.

SAMPLE:

UNIT	SYMBOL
meter	m
gram	g
Newton	N (named after Sir Isaac Newton)
degree Celsius	°C (named after Anders Celsius)

EXCEPTION: The symbol for liter is L. This is used to distinguish it from the number one (1).

2. Prefixes are written with lowercase letters.

SAMPLE:

PREFIX	UNIT	SYMBOL
centi	meter	cm
milli	gram	mg

EXCEPTIONS:

PREFIX	UNIT	SYMBOL
tera	meter	Tm (used to distinguish it from the metric ton, t)
giga	meter	Gm (used to distinguish it from gram, g)
mega	gram	Mg (used to distinguish it from milli, m)

3. Periods are not used in the symbols. Symbols for units are the same in the singular and the plural (no "s" is added to indicate a plural).

   SAMPLE: 1 mm *not* 1 mm. 3 mm *not* 3 mms

4. When referring to a unit of measurement, symbols are not used. The symbol is used only when a number is associated with it.

   SAMPLE: The length of the room is expressed in meters.

   not

   The length of the room is expressed in m. (The *length of the room is* 25 m is correct.)

5. When writing measurements that are less than one, a zero is written before the decimal point.

   SAMPLE: 0.25 m *not* .25 m

6. Separate the digits in groups of three, using commas to the left of the decimal point but not to the right.

   SAMPLE: 5,179,232 mm *not* 5 179 232 mm

   0.56623 mg *not* 0.566 23 mg

   1,346.0987 L *not* 1 346.098 7 L

   A space is left between the digits and the unit of measure.

   SAMPLE: 5,179,232 mm *not* 5,179,232mm

7. Symbols for area measure and volume measure are written with exponents.

   SAMPLE: 3 cm^2 *not* 3 sq cm 4 km^3 *not* 4 cu km

8. Metric words with prefixes are accented on the first syllable. In particular, kilometer is pronounced "kill′-o-meter." This avoids confusion with words for measuring devices that are generally accented on the second syllable, such as thermometer (ther-mom′-e-ter).

Metric Relationships

The base units in SI metrics include the meter and the gram. Other units of measure are related to these units. The relationship between the units is based on powers of ten and uses these prefixes:

kilo (1,000) hecto (100) deka (10) deci (0.1) centi (0.01) milli (0.001)

These tables show the most frequently used units with an asterisk (*).

Metric Length

10 millimeters (mm)*	=	1 centimeter (cm)*
10 centimeters (cm)	=	1 decimeter (dm)
10 decimeters (dm)	=	1 meter (m)*
10 meters (m)	=	1 dekameter (dam)
10 dekameters (dam)	=	1 hectometer (hm)
10 hectometers (hm)	=	1 kilometer (km)*

- To express a metric length unit as a smaller metric length unit, multiply by a positive power of ten such as 10, 100, 1,000, 10,000, etc.
- To express a metric length unit as a larger metric length unit, multiply by a negative power of ten such as 0.1, 0.01, 0.001, 0.0001, etc.

Metric Area Measure

100 square millimeters (mm^2)	=	1 square centimeter (cm^2)*
100 square centimeters (cm^2)	=	1 square decimeter (dm^2)
100 square decimeters (dm^2)	=	1 square meter (m^2)
100 square meters (m^2)	=	1 square dekameter (dam^2)
100 square dekameters (dam^2)	=	1 square hectometer (hm^2)
100 square hectometers (hm^2)	=	1 square kilometer (km^2)

- To express a metric area unit as a smaller metric area unit, multiply by 100, 10,000, 1,000,000, etc.
- To express a metric area unit as a larger metric area unit, multiply by 0.01, 0.0001, 0.000001, etc.

Metric Volume Measure for Solids

1,000 cubic millimeters (mm^3)	=	1 cubic centimeter (cm^3)*
1,000 cubic centimeters (cm^3)	=	1 cubic decimeter (dm^3)
1,000 cubic decimeters (dm^3)	=	1 cubic meter (m^3)
1,000 cubic meters (m^3)	=	1 cubic dekameter (dam^3)
1,000 cubic dekameters (dam^3)	=	1 cubic hectometer (hm^3)
1,000 cubic hectometers (hm^3)	=	1 cubic kilometer (km^3)

- To express a metric volume unit for solids as a smaller metric volume unit for solids, multiply by 1,000, 1,000,000, 1,000,000,000, etc.
- To express a metric volume unit for solids as a larger metric volume unit for solids, multiply by 0.001, 0.000001, 0.000000001, etc.

Metric Volume Measure for Fluids

10 milliliters (mL)*	=	1 centiliter (cL)
10 centiliters (cL)	=	1 deciliter (dL)
10 deciliters (dL)	=	1 liter (L)*
10 liters (L)	=	1 dekaliter (daL)
10 dekaliters	=	1 hectoliters (hL)
10 hectoliters (hL)	=	1 kiloliter (kL)

- To express a metric volume unit for fluids as a smaller metric volume unit for fluids, multiply by 10, 100, 1,000, 10,000, etc.
- To express a metric volume unit for fluids as a larger metric volume unit for fluids, multiply by 0.1, 0.01, 0.001, 0.0001, etc.

Metric Volume Measure Equivalents

1 cubic decimeter (dm^3)	=	1 liter (L)
1,000 cubic centimeters (cm^3)	=	1 liter (L)
1 cubic centimeter (cm^3)	=	1 milliliter (mL)

Metric Mass Measure

10 milligrams (mg)*	=	1 centigram (cg)
10 centigrams (cg)	=	1 decigram (dg)
10 decigrams (dg)	=	1 gram (g)*
10 grams (g)	=	1 dekagram (dag)
10 dekagrams (dag)	=	1 hectogram (hg)
10 hectograms (hg)	=	1 kilogram (kg)*
10 kilograms (kg)	=	1 megagram (Mg)*

- To express a metric mass unit as a smaller metric mass unit, multiply by 10, 100, 1,000, 10,000, etc.
- To express a metric mass unit as a larger metric mass unit, multiply by 0.1, 0.01, 0.001, 0.0001, etc.

Metric measurements are expressed in decimal parts of a whole number. For example, one-half millimeter is written as 0.5 mm.

In calculating with the metric system, all measurements are expressed using the same prefixes. If answers are needed in millimeters, all parts of the problem should be expressed in millimeters before the final solution is attempted. Diagrams that have dimensions in different prefixes must first be expressed using the same unit.

English-Metric Equivalents

Length Measure

1 inch (in)	=	25.4 millimeters (mm)
1 inch (in)	=	2.54 centimeters (cm)
1 foot (ft)	=	0.308 meter (m)
1 yard (yd)	=	0.9144 meter (m)
1 mile (mi)	≈	1.609 kilometers (km)
1 millimeter (mm)	≈	0.03937 inch (in)
1 centimeter (cm)	≈	0.39370 inch (in)
1 meter (m)	≈	3.28084 feet (ft)
1 meter (m)	≈	1.09361 yards (yd)
1 kilometer (km)	≈	0.62137 mile (mi)

Area Measure

1 square inch (sq in)	=	645.16 square millimeters (mm^2)
1 square inch (sq in)	=	6.4516 square centimeters (cm^2)
1 square foot (sq ft)	≈	0.092903 square meter (m^2)
1 square yard (sq yd)	≈	0.836127 square meter (m^2)
1 square millimeter (mm^2)	≈	0.001550 square inch (sq in)
1 square centimeter (cm^2)	≈	0.15500 square inch (sq in)
1 square meter (m^2)	≈	10.763910 square feet (sq ft)
1 square meter (m^2)	≈	1.19599 square yards (sq yd)

Volume Measure for Solids

1 cubic inch (cu in)	=	16.387064 cubic centimeters (cm^3)
1 cubic foot (cu ft)	≈	0.028317 cubic meter (m^3)
1 cubic yard (cu yd)	≈	0.764555 cubic meter(m^3)
1 cubic centimeter (cm^3)	≈	0.061024 cubic inch (cu in)
1 cubic meter (m^3)	≈	35.314667 cubic feet (cu ft)
1 cubic meter (m^3)	≈	1.307951 cubic yards (cu yd)

Volume Measure for Fluids

1 gallon (gal)	≈	3,785.411 cubic centimeters (cm^3)
1 gallon (gal)	≈	3.785411 liters (L)
1 quart (qt)	≈	0.946353 liter (L)
1 ounce (oz)	≈	29.573530 cubic centimeters (cm^3)
1 cubic centimeter (cm^3)	≈	0.000264 gallon (gal)
1 liter (L)	≈	0.264172 gallon (gal)
1 liter (L)	≈	1.056688 quarts (qt)
1 cubic centimeter (cm^3)	≈	0.033814 ounce (oz)

Mass Measure

1 pound (lb)	≈	0.453592 kilogram (kg)
1 pound (lb)	≈	453.59237 grams (g)
1 ounce (oz)	≈	28.349523 grams (g)
1 ounce (oz)	≈	0.028350 kilogram (kg)
1 kilogram (kg)	≈	2.204623 pounds (lb)
1 gram (g)	≈	0.002205 pound (lb)
1 kilogram (kg)	≈	35.273962 ounces (oz)
1 gram (g)	≈	0.035274 ounce (oz)

Section 3

Formulas

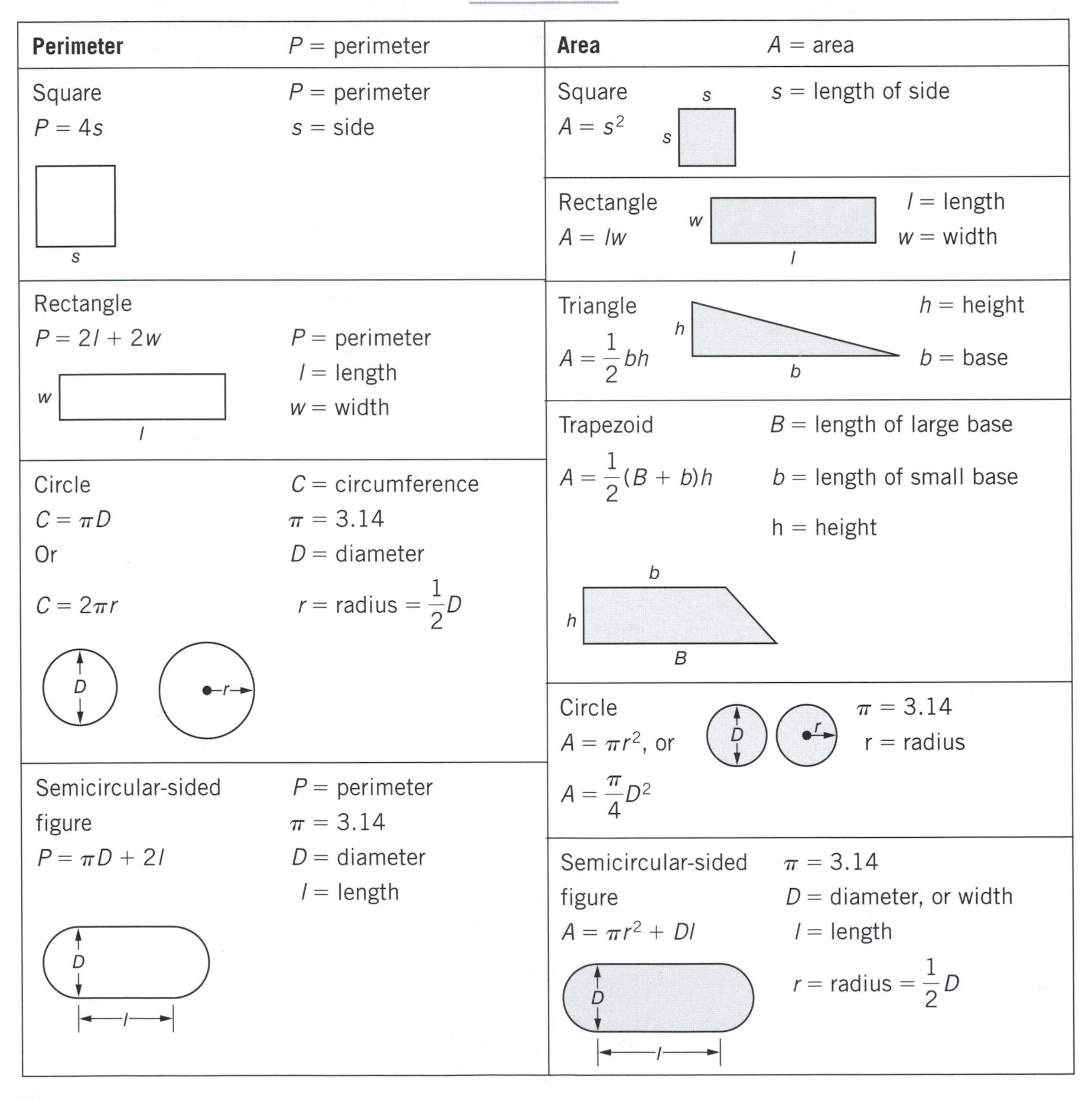

Perimeter	P = perimeter
Square $P = 4s$	P = perimeter s = side
Rectangle $P = 2l + 2w$	P = perimeter l = length w = width
Circle $C = \pi D$ Or $C = 2\pi r$	C = circumference $\pi = 3.14$ D = diameter r = radius $= \frac{1}{2}D$
Semicircular-sided figure $P = \pi D + 2l$	P = perimeter $\pi = 3.14$ D = diameter l = length

Area	A = area
Square $A = s^2$	s = length of side
Rectangle $A = lw$	l = length w = width
Triangle $A = \frac{1}{2}bh$	h = height b = base
Trapezoid $A = \frac{1}{2}(B + b)h$	B = length of large base b = length of small base h = height
Circle $A = \pi r^2$, or $A = \frac{\pi}{4}D^2$	$\pi = 3.14$ r = radius
Semicircular-sided figure $A = \pi r^2 + Dl$	$\pi = 3.14$ D = diameter, or width l = length r = radius $= \frac{1}{2}D$

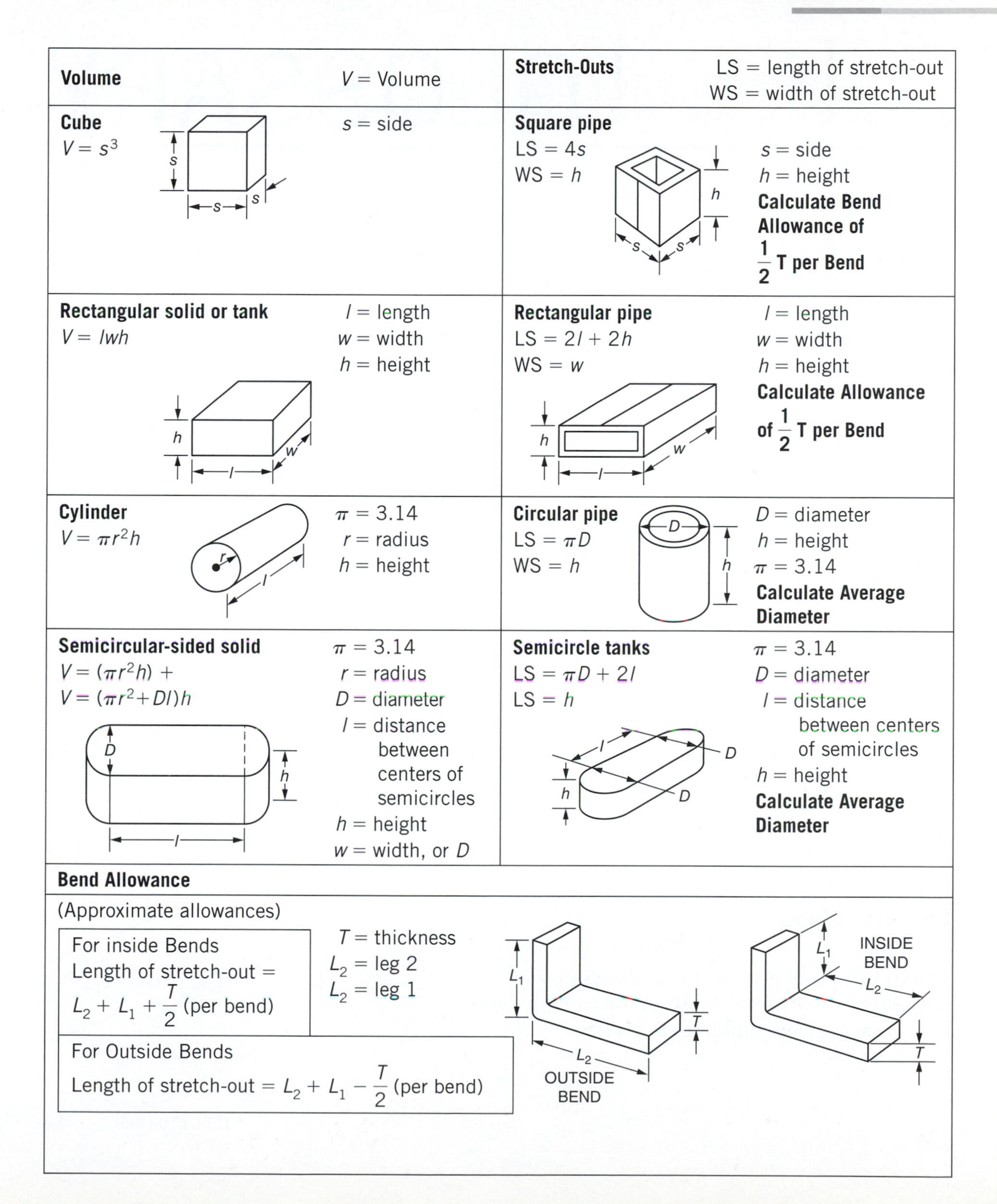

Volume V = Volume	**Stretch-Outs** LS = length of stretch-out WS = width of stretch-out
Cube $V = s^3$ s = side	**Square pipe** LS = $4s$ WS = h s = side h = height **Calculate Bend Allowance of $\frac{1}{2}$ T per Bend**
Rectangular solid or tank $V = lwh$ l = length w = width h = height	**Rectangular pipe** LS = $2l + 2h$ WS = w l = length w = width h = height **Calculate Allowance of $\frac{1}{2}$ T per Bend**
Cylinder $V = \pi r^2 h$ $\pi = 3.14$ r = radius h = height	**Circular pipe** LS = πD WS = h D = diameter h = height $\pi = 3.14$ **Calculate Average Diameter**
Semicircular-sided solid $V = (\pi r^2 h) +$ $V = (\pi r^2 + Dl)h$ $\pi = 3.14$ r = radius D = diameter l = distance between centers of semicircles h = height w = width, or D	**Semicircle tanks** LS = $\pi D + 2l$ LS = h $\pi = 3.14$ D = diameter l = distance between centers of semicircles h = height **Calculate Average Diameter**

Bend Allowance

(Approximate allowances)

For inside Bends
Length of stretch-out = $L_2 + L_1 + \frac{T}{2}$ (per bend)

T = thickness
L_2 = leg 2
L_2 = leg 1

For Outside Bends
Length of stretch-out = $L_2 + L_1 - \frac{T}{2}$ (per bend)

GLOSSARY

Acetylene, Oxy-acetylene Torch Acetylene is a highly combustible gas made of carbon and hydrogen, used as a fuel gas in the oxy-acetylene cutting or welding process. The temperature of the acetylene flame alone could not do cutting or welding; in combination with a forced jet of oxygen, however, the temperature is raised considerably so that those processes are accomplished.

Alloy Fusing of molten metal atoms such as iron with another material, such as carbon, that changes the properties of the original material. The new properties may include increased strength, flexibility, resistance to rusting or corrosion, etc. Various additives including nickel, molybdenum, manganese, etc., may add other changes to the alloy. Any change may or may not be acceptable depending on the use needed of the alloy.

Angle Steel rolled and formed with a cross-section shaped like the capital letter L. The sides, or legs, of steel angle can be of equal length, i.e., 1″ × 1″, 3″ × 3″, or can be of unequal length, i.e., 2″ × 3″, 3″ × 5″. Angle is used for forming the joints in girders, boilers, etc.

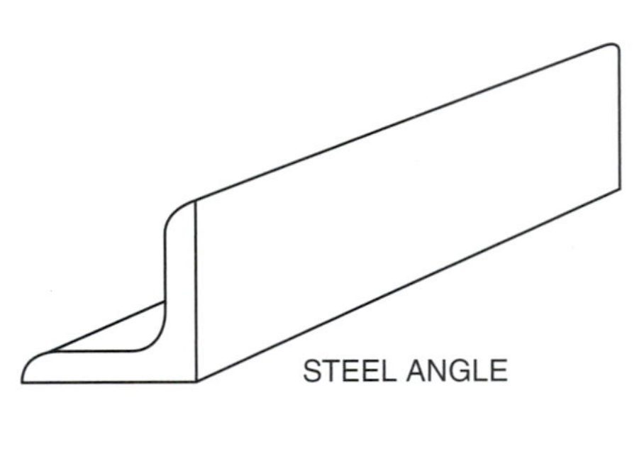

Channel A flanged steel beam whose three sections are composed of a web (the center section), with two equally sized legs. Channel, sometimes called C-channel, can be used as structural components in items ranging from buildings to trailers.

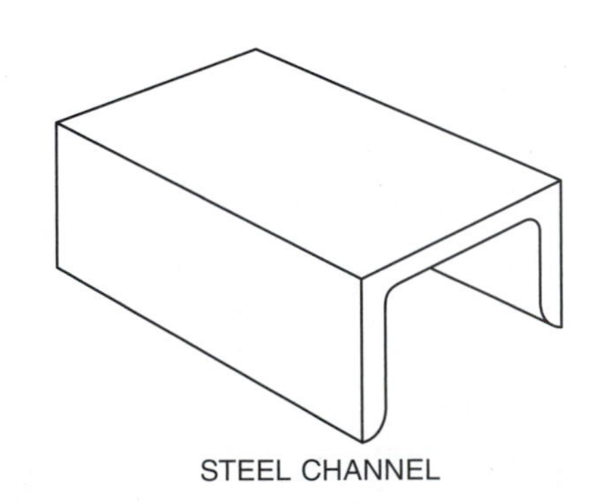

Cold-rolled steel Steel that is rolled from a cold bar of steel. Cold rolling produces steel to closer tolerances than hot rolling, and has a shiny, nickel-like appearance.

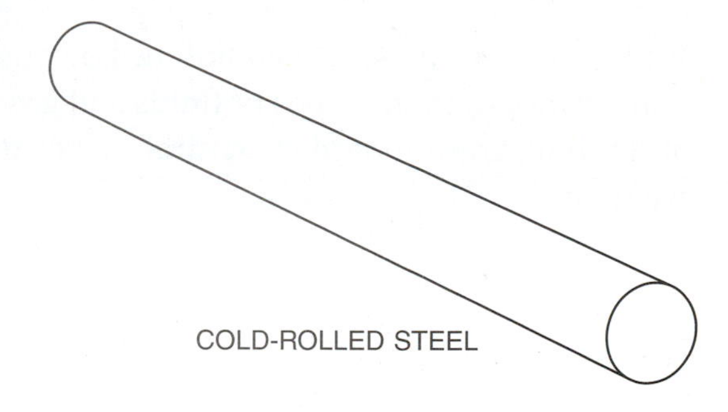

Flame-cutting The cutting of steel using an oxygen-fuel torch, which is combined with the cutting ability of an oxygen jet.

Flat bar Flat bar is furnished in different widths and thicknesses, usually in 10-, 12-, or 20-foot lengths. Some companies manufacture the material up to 6 inches in width, and others manufacture it in 8-inch or 10-inch widths.

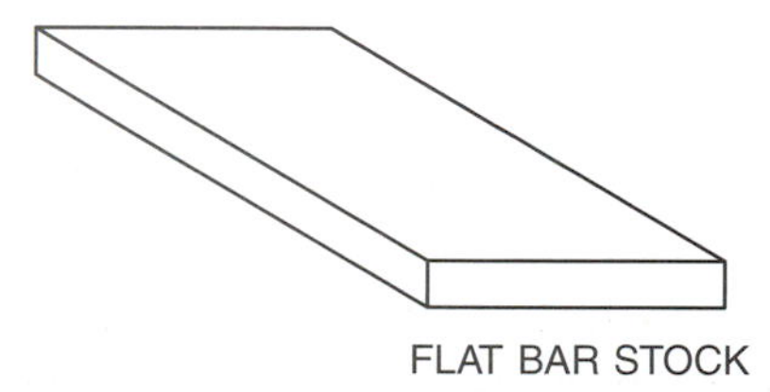

Hot-rolled steel Steel that is rolled or milled while glowing hot. Hot-rolled steel is generally slightly oversized, and has a dark mill scale on its surface.

I beam A beam that has a cross section shaped like the letter I. The I beam is composed of a center web with equally sized flanges on both ends.

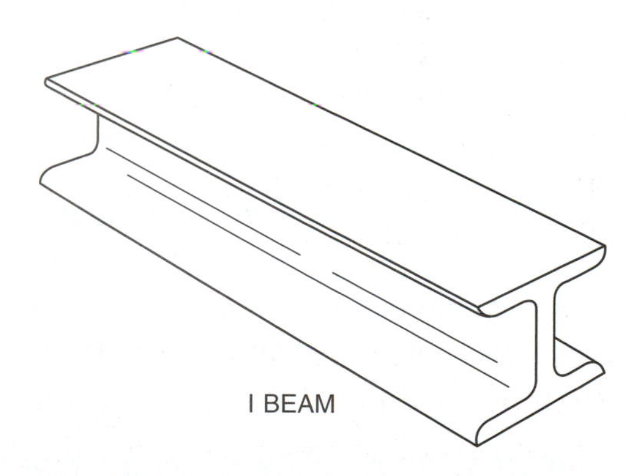

Kerf Material waste caused by the cutting device used, such as a torch, saw blade, or grind wheel.

Mild steel As a rule, a steel with a carbon range between 0.05–0.30 percent is called low-carbon steel, or mild steel. Mild steel is the most commonly found material used in products made from steel.

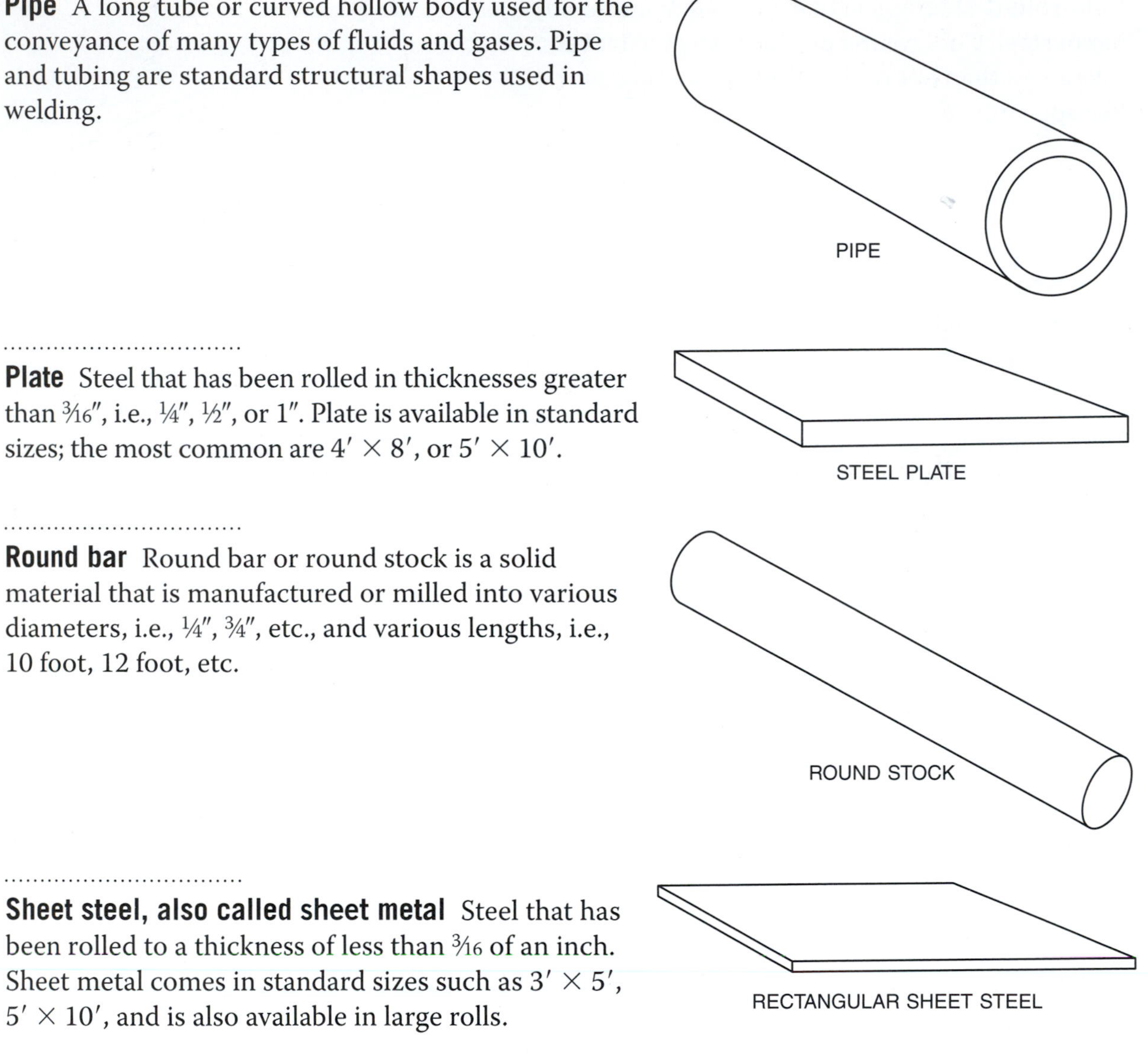

Pipe A long tube or curved hollow body used for the conveyance of many types of fluids and gases. Pipe and tubing are standard structural shapes used in welding.

Plate Steel that has been rolled in thicknesses greater than 3/16″, i.e., 1/4″, 1/2″, or 1″. Plate is available in standard sizes; the most common are 4′ × 8′, or 5′ × 10′.

Round bar Round bar or round stock is a solid material that is manufactured or milled into various diameters, i.e., 1/4″, 3/4″, etc., and various lengths, i.e., 10 foot, 12 foot, etc.

Sheet steel, also called sheet metal Steel that has been rolled to a thickness of less than 3/16 of an inch. Sheet metal comes in standard sizes such as 3′ × 5′, 5′ × 10′, and is also available in large rolls.

Square tube A hollow structural tube which is generally shaped either as a square or a rectangle. Square tubing is used for many purposes in welding.

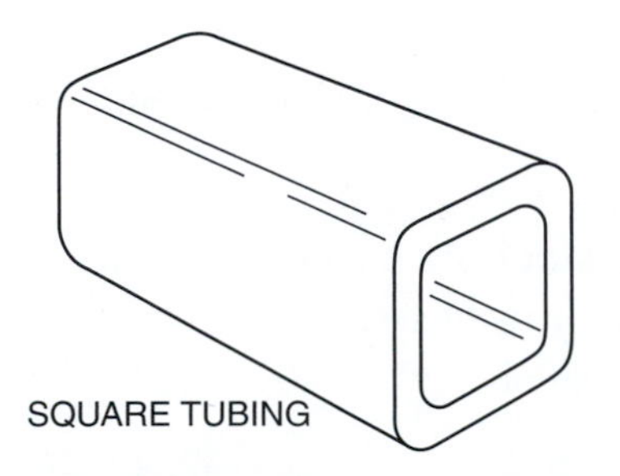

Wide-flange beam Steel formed in the shape of an I beam, but with wider flanges. The cross sections resemble the capital letter "H."

WIDE-FLANGE BEAM

Properties and Characteristics of Metals and Materials

Compressive strength A materials' resistance to compressive force without breakage or deformation.

(DIRECTION OF ARROWS = DIRECTION OF FORCE)

Conductivity Capability of travel in a material by electricity, heat, vibration, or other energetic impulses.

Ductility Capability of a material to be worked by pulling or stretching (i.e., wire), without breakage or rupture (material failure).

Grain Granular/crystalline internal structure of a metal formed as it solidifies/cools from the molten/liquid/heated state. The direction of the granular/crystalline structure, grain size, and speed/duration of cooling (rapid vs. slow) determines various types of strength or failure point weaknesses.

Liquid metal embrittlement A loss of ductility, which can lead to stress fracture (failure) at the weldpoint where a superheated metal (molten state) contacts cooler metal (solid state). (Also see Grain.)

Lustre Capability of a material to accept a shine on its surface.

Malleability Capability of a material to be shaped or worked, by pressure or hammer-blows, without breakage or rupture.

Shear strength A materials' resistance to breaking or rupture when stress is applied in a 90° deflection to length.

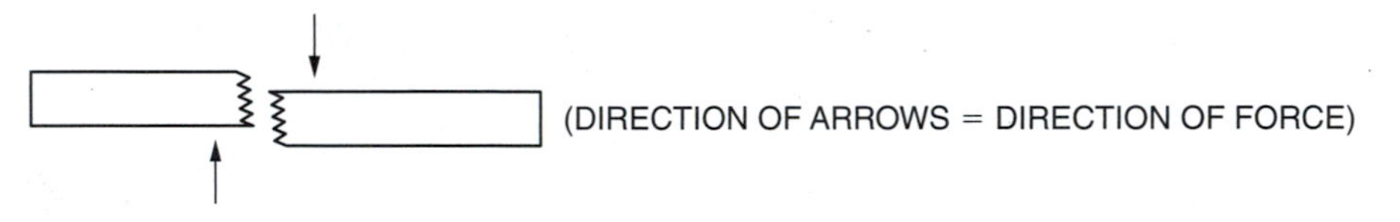

Tensile strength A materials' resistance to breaking or rupture when stress is applied, as being pulled apart in opposite directions (lengthwise).

Toughness/hardness A materials' resistance to fracture or rupture when stressed; also, its resistance to abrasion or cutting, i.e., threading.

Yield strength/yield point A materials' point, when stress is applied, where elastic deformation (temporary shape change) advances to plastic deformation (permanent shape change). At the elastic deformation stage the material will return to its' original shape, while at the plastic deformation stage the change is permanent.

ANSWERS
TO ODD-NUMBERED PROBLEMS

Section 1
Whole Numbers

Unit 1
Addition of Whole Numbers

1. 378 ft
3. 2,031 lbs

Unit 2
Subtraction of Whole Numbers

1. 15″
3. 6″

Unit 3
Multiplication of Whole Numbers

1. a. 272″
   b. 442″
   c. 476 lbs
3. 9,882″
5. a. 1,800″
   b. 14,400″

Unit 4
Division of Whole Numbers

1. 5 pieces
3. 12 hours
5. a. 200 lbs
   b. $75.00 each

Section 2
Common Fractions

Unit 5
Introduction to Common Fractions

Fractions

1. a. $\frac{3}{12}' = \frac{1}{4}'$
   b. $\frac{5}{12}'$
   c. $\frac{8}{12}' = \frac{2}{3}'$
   d. $\frac{11}{12}'$
3. a. $\frac{1}{8}''$
   b. $\frac{3}{8}''$
   c. $\frac{4}{8}'' = \frac{1}{2}''$
   d. $\frac{6}{8}'' = \frac{3}{4}''$

Improper Fractions: Exercises

1. $\frac{4}{3} = 1\frac{1}{3}$
3. $\frac{16}{16} = 1$

Unit 6
Measuring Instruments: The Tape Measure, Caliper, and Micrometer

Exercises

1. $B = \frac{2}{8}'' = \frac{1}{4}''$
   $C = \frac{5}{8}$
   $D = 1\frac{1}{8}''$
   $E = 2\frac{1}{8}''$
3. $A = 3'9\frac{3}{4}''$
   $B = 8'1\frac{1}{4}''$

Unit 7
Addition of Common Fractions

Exercises

1. 5/16
3. 10 1/16

Practical Problems

1. 7 5/8″
3. 27 3/8 lbs
5. 7 15/16″
7. 1 7/8″
9. 19 1/8″

Unit 8
Subtraction of Common Fractions

1. 8 1/8″
3. a. 7 3/6″
   b. 6 15/16
5. 3 7/8″
7. 3 9/16″

Unit 9
Multiplication of Common Fractions

Exercises

1. 29 5/16

Practical Problems

1. 32 1/2″
3. 56 7/16″
5. 21 3/4″
7. 111 3/8 rods

Unit 10
Division of Common Fractions

Exercises

1. 1 1/5
3. 3 9/74

Practical Problems

1. a. 6 pieces
   b. 6 pieces
3. 3 5/8″
5. a. 32 pieces
   b. 1/4″ scrap
7. 3 pieces

Unit 11
Combined Operations with Common Fractions

1. 29 13/16″
3. a. 42 3/8″
   b. 8″

Section 3
Decimal Fractions

Unit 12
Introduction to Decimal Fractions and Rounding Numbers

1. a. 0.3 mile
   b. 0.5 mile
   c. 0.8 mile
3. a. 8
   b. 12
   c. 10
   d. 17
5. 26.05

Calculator Exercises

a. 24
b. 397
c. 144.3′
d. 8.875″
e. 0.625
f. 33
g. 8.754 cm
h. 0.007
i. 1,432.1 mi
j. 1,875′
k. 52.6
l. 1.735
m. 15.573
n. 5
o. 0.2
p. 17.5
q. 681.25

Calculator Formula Exercises

r. 2,600.125 square feet
s. 671.62 square inches

Unit 13
Addition and Subtraction of Decimal Fractions

1. a. 30.74
   b. 17.734
   c. 98.05
   d. 4.732
   e. 5.315
   f. 22.828
3. 36.675 lbs

Unit 14
Multiplication of Decimals

Exercises

1. 54

Practical Problems

1. 308.75″
3. 60.42 ft^3
5. 36.4 lbs
7. 4.875 lbs
9. 26 lbs

Unit 15
Division of Decimals

1. a. .66″
   b. .40″
3. 26 pieces
5. a. 8.41″
   b. 6 strips
   c. .75″

Unit 16
Decimal Fractions and Common Fraction Equivalents

1. a. 4.0625″
   b. 6.25″
   c. .375″
   d. = 5⁄8″
   e. = 3⁄4″
3. a. 10 1⁄2″
   b. 2 3⁄4″
   c. 3 1⁄4″
   d. 3⁄4″
5. a. 12⁄32″ = 3⁄8″
   b. 1⁄16″
   c. 15⁄16″
7. a. 2.25″
   b. 0.1875″
   c. 6.5625″
   d. 3.3594″
   e. 11.875″
   f. 3.6875″
   g. 4.1875″
   h. 0.625″

Unit 17
Tolerances

1. a. 0.755
   b. 0.745
3. a. 3′ 6.35″
   b. 3′ 6.23″
5. a. 18′ 4 9⁄16″
   b. 18′ 4 7⁄16″
7. a. 10 21⁄32
   b. 10 19⁄32

Unit 18
Combined Operations with Decimal Fractions

1. 6.3125″
3. 302.25″
5. a. 9 supports
   b. 8 supports
7. flange: a. 9.37″
   b. 9.315″
   bolt-hole circle: a. 3.254″
   b. 3.246″
9. 0.39″

Unit 19
Equivalent Measurements

1. 276″
3. 2′
5. 4.5′
7. 38.75″

Section 4
Averages, Percentages, and Multipliers

Unit 20
Averages

1. 67.5 miles
3. 40 lbs
5. 1.297″
7. 35 lbs

Unit 21
Percents and Percentages (%)

1. a. 0.16
   b. 0.05
   c. 0.008
   d. 0.605
   e. 0.2325
   f. 1.25
   g. 2.20
3. 361.61 in^2
5. a. 96.25%
   b. 85%
   c. 100%

(*continued*)

7.

	Fraction	Decimal	%
a.	3/8	0.375	37.5%
b.	4/5	0.80	80%
c.	2 1/8	2.125	212.5%
d.	7/32	0.21875	21.875%
e.	3/4	0.75	75%
f.	16/16	1.0	100%
g.	any, i.e. 2/2, 3/3 etc.	1.0	100%

9. $247.56

Section 5
Metric System Measurements

Unit 22
The Metric System of Measurements

1.

Millimeters	Centimeters
A = 10	1.0
B = 17	1.7
C = 27	2.7
D = 32	3.2
E = 45	4.5
F = 53	5.3
G = 77	7.7
H = 99	9.9

3. 100 cm
5. D. 102.4 cm
7. C. 291 mm
9. a. 16.51 cm
   b. 1,000 mm
   c. 60.96 cm
11. 11.25 m

Unit 23
English-Metric Equivalent Unit Conversions

1. 2.499 m
3. a. 70.87″
   b. 5.71″
5. a. 3.359 m
   b. 146.05 mm
   c. 92.075 mm
   d. 1.181 m

Unit 24
Combined Operations with Equivalent Units

1. 14″
3. 0.15
5. a. 144.46 cm
   b. 103.19 cm
   1031.88 mm
7. a. 5410.2 cm
   b. 3124.2 cm
   c. 64.2 m

Section 6
Computing
Geometric Measure and Shapes

Unit 25
Perimeter of Squares and Rectangles, Order of Operations

1. a. 7″
   b. 64 cm
   c. 774.7 mm
   d. 38″
   e. 3.12 m

Unit 26
Area of Squares and Rectangles

1. 39 1/16 in^2
3. 144 in^2
5. 144 in^2
7. C

Unit 27
Area of Triangles and Trapezoids

1. 32 in^2
3. 72 in^2
5. 1,227.42 cm^2
7. 55 in^2
9. 27 in^2
11. 562.5 ft^2

Unit 28
Volume of Cubes and Rectangular Shapes

1. 1,728 in^3
   1 ft^3
3. 2.949 ft^3
5. 18.75 in^3
7. 324 in^3
9. .35 m^3

Unit 29
Volume of Rectangular Containers

1. 6.381 gallons
3. 134.649 gallons
5. 12,167 in^3
7. 1,116.8 in^3
9. No
11. a. 691,200 in^3
    b. 400 ft^3
13. 52.0″
15. 166.35 liters

Unit 30
Circumference of Circles, and Perimeter of Semicircular-Shaped Figures

1. A. 30.35′
      364.24″
   B. 13.345′ or 13.35′
      160.14″
3. 24,849.96 miles
5. 24.018 cm

Unit 31
Area of Circular and Semicircular Figures

1. 1017.36 in^2
3. a. 6,273 in^2
   b. 3,997.665 in^2
   c. 2,275.335 or 2,275.34 in^2
5. 1,160.90 in^2

Unit 32
Volume of Cylindrical Shapes

1. 4,710 in^3
3. 3,102.86 in^3
5. 12.56 ft^3
7. 1.57 ft^3
9. 1,9216.8 in^3
11. No, tank B is 8 times bigger than tank A.

Unit 33
Volume of Cylindrical and Complex Containers

1. a. 4,768.875 in^3
      or 4,768.88 in^3
   b. 5,086.8 in^3
   c. 5,913.405
      or 5,913.41 in^3
3. 115.50 ft^3
5. 174.9 gallons
7. 1,109.0 gallons

Unit 34
Mass (Weight) Measure

1. 280.41 lbs
3. a. 320.47 lbs
   b. 87.77 lbs
5. a. 27.581747 rounds to 27.58 kg
   b. 441.31 kg (developed from 27.581747)
7. 373.16 lbs
9. 2,147.24 lbs

Section 7
Angular Development and Measurement

Unit 35
Angle Development

1. 120°
3. 300°
5. ⅙
7. ¼
9. 4/9
11. 45°
13. a. 45°
    b. 90°
    c. $51\frac{3}{7}$° or 51.43°
15. c.
17. A. 12.56 fps
       8.56 mph
    B. 25.12 fps
       17.13 mph

Unit 36
Angular Measurement

1. 112° 14' 8"
3. 271° 2' 19"
5. 80° 35' 38"
7. 32° 36' 38"
9. 180° 24"
11. 330°
13. 137° 30'
15. 68° 45'

Unit 37
Protractors

1. a.

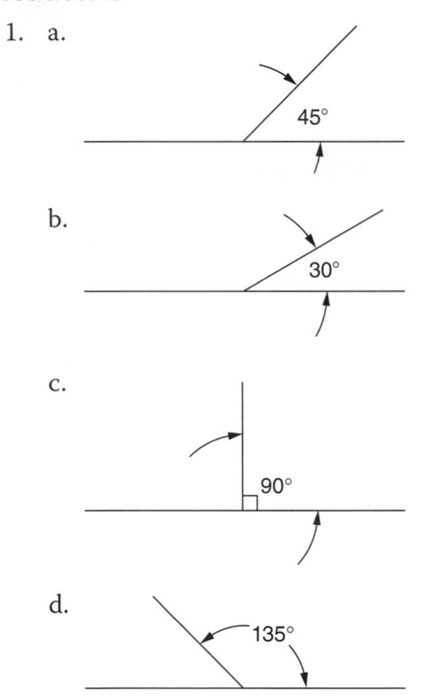

e.

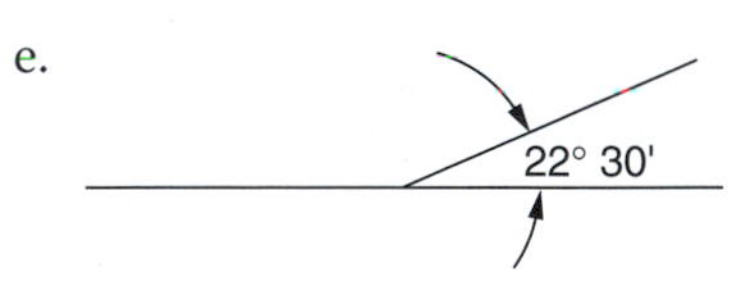

Section 8
Bends, Stretchouts, Economic Layout and Take-Offs

Unit 38
Bends and Stretchouts of Angular Shapes

1. 169.20 cm
3. 74¾″ × 63″
5. 8′ 8⅛″

Unit 39
Bends and Stretchouts of Circular and Semicircular Shapes

1. 126¾″ × 63¾″
3. a. 180.94″
   b. 19⅞″
   c. 79 7/16″
5. a. 24.53″
   b. 18.25″

Unit 40
Economical Layouts of Rectangular Plates

1. a. 72 pieces
   b. 63
3. a. 8 pieces
   b. 2″ × 6½″
5. 70
7. 40
9. 54
11. a. 54
    b. 52 pieces

Unit 41
Economic Layout of Odd-Shaped Pieces; Take-Offs

1. a. 6 circles
   b. 2
3. a. 3 lengths
   b. 19.68%
5. 9 pieces
7. 205 pieces
9. a. 103.22″
   b. 3′ 9¼″
   c. 34.64″